Bastian Hartmann

Human Worker Activity Recognition in Industrial Environments

Human Worker Activity Recognition in Industrial Environments

by
Bastian Hartmann

KIT Scientific Publishing

Dissertation, Karlsruher Institut für Technologie
Fakultät für Elektrotechnik und Informationstechnik, 2011

Impressum

Karlsruher Institut für Technologie (KIT)
KIT Scientific Publishing
Straße am Forum 2
D-76131 Karlsruhe
www.ksp.kit.edu

KIT – Universität des Landes Baden-Württemberg und nationales
Forschungszentrum in der Helmholtz-Gemeinschaft

KIT Scientific Publishing 2011
Print on Demand

ISBN 978-3-86644-643-4

Human Worker Activity Recognition in Industrial Environments

Zur Erlangung des akademischen Grades eines

Doktor-Ingenieurs

von der Fakultät für

Elektrotechnik und Informationstechnik

des Karlsruher Institut für Technologie (KIT)

genehmigte

Dissertation

von

Dipl.-Ing. Bastian Hartmann

geb. in Heidelberg

Tag der mündlichen Prüfung:	03. 02. 2011
Hauptreferent:	Prof. Dr.-Ing. Gert F. Trommer
1. Korreferent:	Prof. Dr. rer. nat. Norbert Link
2. Korreferent:	Prof. Dr.-Ing. Klaus D. Müller-Glaser

Karlsruhe 2011

Vorwort

Die vorliegende Arbeit entstand während meiner Tätigkeit als wissenschaftlicher Mitarbeiter am Institut für Angewandte Forschung der Hochschule Karlsruhe.

Meinem Doktorvater, Herrn Professor Dr.-Ing. Gert F. Trommer, möchte ich ganz besonders herzlich für die Übernahme des Hauptreferats und für seine Hilfe bei dieser Arbeit durch wertvolle Anregungen und Diskussionen danken. Des Weiteren danke ich Herrn Professor Dr. rer. nat. Norbert Link für die Übernahme des ersten Korreferats, sowie vor allem für die Unterstützung meiner Ideen und für die hilfreichen Diskussionen. Darüber hinaus danke ich auch Herrn Professor Dr.-Ing. Klaus D. Müller-Glaser für die Übernahme des zweiten Korreferats und das Interesse an meiner Arbeit.

Auch möchte ich allen Kollegen und Mitarbeitern am Institut danken, die für eine kollegiale Arbeitsatmosphäre gesorgt haben und mich während meiner Arbeit am Institut unterstützt haben.

Insbesondere danke ich auch Frau Dr. Gieta Dewal und Herrn Dr. Ingo Schwab für die Durchsicht des Manuskripts und die vielen hilfreichen Ratschläge.

Mein Dank gilt auch allen Studenten, die durch ihre fleißige Arbeit wichtige Beiträge geliefert haben.

Karlsruhe, im Januar 2011

Bastian Hartmann

Zusammenfassung

Moderne Produktionsanlagen sind geprägt von einer Vielzahl automatisierter Prozesse. Trotz des hohen Grades an Automatisierung existieren hingegen auch viele Fälle in denen der Einsatz von menschlichen Arbeitskräften aufgrund ihrer Flexibilität gegenüber maschinellen Produktionskomponenten von Vorteil ist. Der Trend in Richtung intelligenter Fabriken, die in der Lage sind die Produktion selbstständig zu steuern, macht es wiederum erforderlich, dass menschliche Arbeiter (auch Werker genannt) in vollautomatisch gesteuerte Fertigungsprozesse nahtlos eingebunden werden können. Die nahtlose Integration von Werkern in vollautomatisch gesteuerte Fertigungsprozesse erfordert eine Mensch-Maschine Schnittstelle, die sowohl Rückmeldungen und Kommandos des Werkers an das Produktionssystem als auch automatisch menschliche Tätigkeiten erkennt.
Die vorliegende Arbeit behandelt eine Mensch-Maschine Schnittstelle, die diese Funktionalitäten bietet und in Hinsicht auf mögliche Anwendungen aus flexiblen und wieder verwendbaren Komponenten besteht.

Die Schätzung der Position beliebiger mit der Werkertätigkeit verbundener Objekte stellt einen Grundbaustein zur Erkennung von Tätigkeiten im vorgestellten Schnittstellenkonzept dar. Das Positionserkennungssystem basiert auf einer Inertialnavigationslösung, die Daten einer inertialen Messeinheit und eines markenbasierten Videotrackingsystems mittels eines erweiterten Kalmanfilters fusioniert. Durch die Kombination dieser Art von Sensoren ist es möglich, die Vorteile der hohen Abtastrate der Inertialsensoren und der absoluten Positioniergenauigkeit des Videotrackingsystems zu verbinden. Des Weiteren erlaubt das Positionserkennungssystem die Kompensation von durch Verdeckungen hervorgerufenen kurzfristigen Ausfällen des Videotrackingsystems und bietet eine gewisse Skalierbarkeit in Bezug auf den Arbeitsraum.

Die automatische Erkennung der Werkertätigkeiten basiert auf einem Konzept, das auf der Modellierung von Aufgaben als Sequenzen aus sowohl sicheren als auch unsicheren Informationen über Handlungsprimitiven beruht. Dazu werden zunächst Ortsinformationen aus Positionsschätzungen und elementare Handlungen aus Inertialsensordaten mittels maschineller Lernverfahren klassifiziert. Aus den Klassifikationsergebnissen werden in einem weiteren Schritt Handlungsprimitiven abgeleitet. Die auf diese Weise gewonnen Informationen über Handlungsprimitiven werden in einem übergeordnetem Prozess unter Einbezug von Kontextwissen verarbeitet, um ausgeführte Werkertätigkeiten zu erkennen.

Eine natürliche und komfortable Kommunikation zwischen Werker und Produktionssystem wird durch Rückmeldungen und Kommandos über Handgesten, die über Beschleunigungssensoren erfasst werden, ermöglicht. Das entwickelte Gestenerkennungsverfahren basiert auf einem elastischen Vergleichsmaß für Zeitreihen und ist somit in der Lage Gesten unabhängig von deren Ausführungsgeschwindigkeit zu erkennen. Ein wichtiger Bestandteil dieses Verfahrens ist die Gewinnung von Vergleichsmustern, die als Repräsentanten für Gestenklassen dienen. Diese erfolgt über ein neuartiges Optimierverfahren, das die Klassentrennbarkeit der Vergleichsmuster maximiert.

Die Funktionsweise der Mensch-Maschine Schnittstelle wird anhand von zwei verschiedenen Szenarien aus dem industriellen Umfeld demonstriert. Das erste Szenario behandelt die Erkennung von Werkertätigkeiten beim Schweißen mit einer Handschweißzange. Ein weiteres Szenario, in dem die Montage eines elektrischen Gerätes behandelt wird, dient zur Demonstration der Erkennung von Handmontagetätigkeiten. Durch die behandelten Szenarien wird die industrielle Anwendungstauglichkeit der vorgestellten Mensch-Maschine Schnittstelle gezeigt, sowie die flexible Verwendung der Komponenten demonstriert.

Table of Contents

List of Symbols

Abbreviations

ADL	Activities of Daily Living
BFS	Brute Force Search
BN	Bayesian Network
BCH code	Bose-Chaudhuri-Hocquenghem-Code
CD	Compact Disc
CV	Cross Validation
DAG	Directed Acyclic Graph
DBN	Dynamic Bayesian Network
DOF	Degrees of Freedom
DTW	Dynamic Time Warping
EKF	Extended Kalman Filter
ES	Evolution Strategy
EU	European Union
FP	False Positive
FSM	Finite State Machine
GPS	Global Positioning System
GT	Ground Truth
GUI	Graphical User Interface
HMM	Hidden Markov Model
IMU	Inertial Measurement Unit
INS	Inertial Navigation System
k-NN	k-Nearest Neighbor algorithm
KF	Kalman Filter
MAP	Maximum A Posteriori

μC	Microcontroller
MEMS	Micro-Electro-Mechanical Systems
ML	Maximum Likelihood
PAA	Piecewise Aggregate Approximation
PC	Personal Computer
PCA	Principal Component Analysis
RFID	Radio Frequency Identification
ToA	Time of Arrival
TP	True Positive
UART	Universal Asynchronous Receiver Transmitter
UML	Unified Modeling Language
VT	Video-Tracking
Wi-Fi	Wireless Fidelity

Latin Symbols

a,b,c,d	quaternion elements
A	HMM transition matrix
$\vec{a}$	acceleration vector
B	HMM observation matrix
$\vec{b}_a$	accelerometer biases
$\vec{b}_\omega$	angular rate biases
c_i	class
C	direction cosine matrix
C	DTW class template
C_{opt}	DTW prototype
$d(\cdot)$	distance
d	dimension
$D_{KL\ 1,2}$	Kullback-Leibler divergence
$DTW(C,T)$	dynamic time warping distance between time series C and T
E	DBN evidence
E	warping path endpoint
$\vec{e}$	eigenvector

f	focal length
f_{target}	target function
$\vec{f}()$	Kalman filter system model function
$\vec{g}$	gravity vector
$\vec{h}()$	Kalman filter measurement model function
H_k	Kalman filter measurement noise matrix
k	parameter of the k-nearest neighbor algorithm
K	camera matrix
K_k	Kalman gain
O	HMM observation sequence
$\vec{p}$	position vector
P_k	Kalman filter system state covariance matrix
$\vec{q}$	quaternion
Q_k	Kalman filter system noise covariance matrix
R	rotation matrix
R_k	Kalman filter measurement noise covariance matrix
(s_x, s_y)	image format
S_n	signal part in sliding window
S	HMM state sequence
$\vec{t}$	translation vector
T	pose transformation matrix
T	DTW test template
u, v	image coordinates
$\vec{u}$	Kalman filter model input vector
$\vec{v}$	velocity vector
$\vec{v}$	Kalman filter model measurement noise
$\vec{w}$	Kalman filter model system noise
W	DTW warping path
(x_c, y_c)	principal point
x, y, z	3D position coordinates
$\vec{x}$	Kalman filter model state vector
$\vec{x}$	feature vector
$\vec{x}'$	extracted features
$\vec{z}$	Kalman filter model measurement vector

Greek Symbols

α	startpoint of prototype subseries
β	endpoint of prototype subseries
γ	skew coefficient
ΔT	sampling interval
θ	pitch angle
λ	HMM
λ	evolution strategy parameter
μ	mean value
μ	evolution strategy parameter
$\vec{\pi}$	HMM initial state probability distribution
$\vec{\sigma}$	rotation vector
Σ	covariance matrix
τ	threshold
ϕ	roll angle
Φ_k	Kalman filter state transition matrix
ψ	yaw angle
$\vec{\Psi}$	attitude vector
$\vec{\omega}$	angular rate vector

1. Introduction

Nowadays, computers are omnipresent and serve as facilitation of our daily work and life. An advancement to the use of common desktop computers is the vision of ubiquitous computing or pervasive computing. The term ubiquitous computing refers mainly to microcomputers and intelligent systems, which are integrated into objects that are used by humans. Ubiquitous systems should be able to assist and facilitate human activities without the need of active usage or even without that the user is aware of ubiquitous devices. For bridging the gap between the world of humans and the world of machines, intelligent human-machine interfaces are required. Therefore, intelligent human-machine interfaces can be seen as important components for ubiquitous computing systems.

In this work, an intelligent human-machine interface for human worker activity recognition in industrial environments is presented. The interface is able to recognize task-related worker activities and provides the capability of human-computer interaction via gestures. By means of the presented human-machine interface, workers can be integrated in software-controlled production or in an intelligent manufacturing system.

1.1. Background and Motivation

Advances in manufacturing have always been associated with improvements in product quality and efficiency of production. In times of free market economy, efficient production systems and a good manufacturing quality are factors of particular importance for manufacturers in order to stay competitive. Since the age of mass production, the efficiency of manufacturing technology has been increased by automation. As a result, modern automated production sites allow fast and cost-effective manufacturing of products in large numbers.

However, automated production is not the one and only solution to efficient production and competitive manufacturing systems. One of the major drawbacks of highly automated manufacturing sites is their inflexibility. This drawback has a crucial influence on efficiency in situations where flexible and reconfigurable manufacturing systems are required. The following conditions are examples for the problem of inflexibility in highly automated manufacturing sites:

- Small lot sizes: Because of the fact that automated production systems are focused on the production of a specific product, they occasion considerable reconfiguration costs, if adaptations are needed for manufacturing a new product. Furthermore, manufacturing of multiple product variants at a time is often not or hardly possible. Therefore, a high degree of automation is only feasible if the product to be manufactured exceeds a certain lot size.
- Product changes: Sometimes it happens that the design of a product has to be changed due to several reasons after going into production. Because sudden product changes require new machine configurations they cause machine idle times, which cost a lot of efficiency.
- Malfunctions in assembly lines: Due to the stiff organization of the production flow in assembly lines, a malfunction in one part of the assembly line causes a breakdown of the complete assembly line until the malfunction is fixed. Thus, larger repairs are very cost-intensive.

For these reasons, finding a trade-off between a high degree of automation and flexibility is important for efficient manufacturing systems.

A lot of research in the field of manufacturing technology goes into the direction of intelligent manufacturing systems, which should be able to provide a solution for flexible manufacturing. Approaches for intelligent manufacturing systems exist for instance in form of agent-based manufacturing systems [67, 93] or holonic manufacturing systems [14]. Based on autonomous and cooperating intelligent production units (also called agents), a distributed and flexible concept for intelligent manufacturing is offered by these approachcs.

An important issue, which has to be regarded for intelligent manufacturing systems, is that not only machines, but also workers are present in factories. Because of their flexibility, human workers play an important role in manufacturing systems of the future. However, a problem which arises in here is that in contrast to machines, workers need particular interfaces, if they should be integrated in intelligent or software-controlled manufacturing. Therefore, human-machine interfaces are needed for intelligent manufacturing systems in the future.
An example for the integration of human workers in an intelligent manufacturing system is given by the EU project XPRESS [23]. The ideas of XPRESS and the general need for the integration of human workers into software-controlled manufacturing have been the main intention for this work. Thus, the presented work demonstrates the integration of human workers into the concept of XPRESS and constitutes a realization for the recognition of human worker activities in industrial environments.

1.2. Objectives

The general objective of this work is to provide a human-machine interface for the integration of human workers in intelligent manufacturing systems. Basically, interfaces are devices or methods that allow interactive communication, which means a bidirectional flow of information.
The flow of information from the manufacturing system to the worker can be realized by means of standard human-computer interface methods, e.g. graphical output on a computer monitor (GUI). Because of that, this direction of communication is not considered as part of this work, although the field of visualization methods comprises also sophisticated techniques, such as augmented reality.
A complex issue is constituted by the realization of the flow of information from the worker to the manufacturing system if seamless integration of human workers in the intelligent manufacturing environment is desired. In this case, information about the status of the task executed by the worker is needed for production control purposes. For this reason, standard human-computer interface methods (e.g. mouse and keyboard) are very impractical, because

they require active reporting from the worker. Furthermore, standard methods for active human-computer interaction are inconvenient for working persons, because they are based on access via computer terminals.
Because of these needs, the goal of this work is to realize an unidirectional human-machine interface with the following capabilities:

- Automatic recognition of performed worker tasks: For the seamless integration of human workers in an intelligent manufacturing system, it is required that the interface is able to provide information about the status of tasks executed by the worker. This incorporates the automatic recognition of activities related to tasks and reasoning about the task status from recognized activities. Furthermore, the choice of adequate sensor systems and signal processing methods for obtaining physical information as a basis for activity recognition is required.
- Gesture recognition for worker feedback: Besides the automatic recognition of executed tasks, it is necessary that a worker has a possibility for active communication with a superordinate framework. This comprises a comfortable method for giving feedback (e.g. issuing commands for navigation through a help menu) without leaving the workplace. Since speech, which is the most common and natural way of human communication, is inconvenient for usage in noisy industrial environments, hand gestures constitute a more adequate communication method. Therefore, a gesture recognition system is needed, which should be based on adequate sensing techniques.
- Application independence by means of flexible and reusable components: Another important aspect in the development of a human-machine interface for intelligent manufacturing is the degree of reusability. Since the manufacturing system should be flexible with respect to the manufacturing task, the interface has to be able to cover different production scenarios. Thus, the developed components should not focus on a particular use case, but have the capability to be adapted to new production scenarios.

1.3. Contributions

Based on the objectives, which have been mentioned in the previous section, an intelligent human-machine interface for worker activity recognition in industrial environments has been developed. By means of this work the following contributions are made:

- Robust and scalable indoor object position estimation: As a basic sensing technique, a system has been developed for 3D position tracking of arbitrary objects in industrial environments. The tracking system utilizes an approach similar to an inertial navigation system, based on marker-based video-tracking, inertial sensors and a Kalman filter for data fusion. By making use of the complimentary properties of video sensors and inertial sensors, the system can be used for precise indoor position tracking on different scales and is able to provide high sampling rates.
- Recognition of worker activities and tasks with a flexible approach: The presented activity recognition approach is based on a concept for designing an interface with reusable components. The approach comprises the recognition of manufacturing tasks with state-based methods, which model the interdependencies of low-level worker activities. For task recognition both certain input information and uncertain input information have been considered. Low-level activities are, in turn, recognized with standard machine learning classifiers from position information, such as signals from wearable sensors that indicate actions. It is demonstrated that the investigated application scenarios are compliant with this concept.
- Gesture recognition for comfortable interaction with a superordinate framework: For the recognition of gestures from sensor signals, an online time series template matching approach is presented. The recognition method utilizes a technique called dynamic time warping and is able to recognize temporarily and spatially distorted time series patterns. Representative templates for classes of gestures are found by means of a novel prototype optimization approach. For the human-machine inter-

face, the method is applied to accelerometer signals in order to recognize hand gestures.

- Sample demonstration of the activity recognition approach by means of representative scenarios: In order to validate the task recognition capability of the human-machine interface, the whole system is tested by means of two realistic manufacturing scenarios in the form of real-time demonstrations. The first scenario is a simple spot-welding scenario of the automotive industry, which finds application within the scope of XPRESS. The second scenario is of higher complexity. It contains the recognition of worker activities during a manual assembly task, in which a worker assembles an electronic device. This scenario is also part of the results of XPRESS.
  Furthermore, the capability of comfortable interaction during task execution is demonstrated by means of hand gesture recognition.

It is not known to the authors that a human-machine interface for worker activity recognition in industrial environments is existing, which offers the complete above mentioned functionality in order to provide the capabilities of section 1.2. Thus, the presented human-machine interface constitutes a novel approach to the integration of human workers into intelligent manufacturing systems.

1.4. Thesis Outline

The thesis is organized as follows.
In Chapter 2 related work about human activity recognition in general is reviewed and use cases for the recognition of human activities in industrial environments are identified. Furthermore, the two scenarios that are used for the evaluation, demonstration and application of the achievements of this work are introduced. Based on the use cases, a conceptual approach for the development of worker activity recognition systems is presented. Finally, an overview about the human-machine interface and its components, which have been developed in this work, is given.
In chapter 3 a robust and scalable system for indoor position estimation of

arbitrary objects is proposed. The position estimation system is based on marker-based video-tracking and uses measurements from inertial sensors such as an extended Kalman filter for increasing the sampling rate and for providing robustness to short outages. The performance of the position estimation system is evaluated in experiments with simulated and real data.

In Chapter 4 techniques for the recognition of low-level activities are presented. The presented activity recognition approach is based on action and location information and utilizes statistical learning methods for the classification of worker activities. All proposed methods of this chapter are evaluated and compared in real data experiments with a PC assembly scenario.

In chapter 5 an approach for task level activity recognition is described. By utilizing a statechart model and a hidden Markov model such as a dynamic Bayesian network, tasks can be recognized not only from certain, but also from uncertain low-level activity results. The recognition performance of the hidden Markov model and the dynamic Bayesian network is evaluated and compared in experiments with the PC assembly scenario.

In Chapter 6 an online gesture recognition approach based on an elastic time series distance measure, called dynamic time warping, is presented. The gesture recognition approach utilizes a novel optimization technique for finding time series templates that aims at maximal class separability. For template optimization an evolution strategy and different target functions measuring class separability are proposed. The optimization method and the gesture recognition performance are analyzed and evaluated with a set of recorded hand gestures.

The functionality of the human-machine interface components is finally demonstrated as a complete realtime system in chapter 7. The demonstration consists of experiments with a spot welding scenario and the PC assembly scenario.

In Chapter 8 the work of the thesis is summarized and concluded.

2. Human Worker Activity Recognition

This chapter introduces the topic of human worker activity recognition and gives an overview about the presented activity recognition approach. At first, related work about human activity recognition applications is reviewed. After that, use cases for activity recognition in industrial environments are suggested and scenarios, which have been chosen for demonstration, are introduced. Furthermore, our solution concept for the worker activity recognition system is proposed, and finally, an overview about the developed human-machine interface is given.

2.1. Related Work

The topic of human activity recognition is related to different fields of research and several scenarios are existing, in which activity recognition applications are demonstrated. However, it has to be mentioned that many of the approaches that are presented in literature are prototype systems and did not yet reach the level of maturity that is necessary for large scale application. This is mainly because of the fact that, due to the diversity of human behavior, the recognition of human activities is a problem of high complexity.

Industry and Manual Assembly

An interesting example for the research on wearable computing in working environments is given by the EU project WearIT@work [63]. Within this project, human workers and experts are assisted in their work by wearable computing devices that are able to track performed activities. Sample demonstrations cover scenarios in the sectors of aircraft maintenance, car production, healthcare and emergency response. The work of [100], which is related to the WearIT@work project, is about activity spotting in industrial environments with wearable sensors. The spotting approach is demonstrated by means of two scenarios. The first scenario is a maintenance scenario, in which activities during the execution of a bicycle repair task are detected. The second scenario is an example for quality assurance in the automotive industry. In this scenario the steps of a car inspection task are recognized. In [30] a vision based system is presented for the recognition of manual assembly tasks. The method of this work is demonstrated by means of a fictive manual assembly scenario, in which screw and put actions are detected. Another approach for the recognition of manual assembly tasks is given in [5]. This work is based on activity recognition with sensors attached to workpieces and is demonstrated by means of a furniture assembly task. In [9] a system is presented for the detection of dangerous situations that may occur when human workers operate industrial machines. The system is demonstrated in an application where accidents with circular saws are avoided.

Medical Applications and ADL Recognition

Besides industrial applications and manual assembly tasks, medical purposes and health care scenarios form an important category for applications of assisting activity recognition systems. In [8] activities of people working in hospitals are recognized in order to provide context information for assistance. Other applications in the medical area deal with activity recognition in homes for elderly care [102, 123] or the detection of alert situations [12]. The recognition of activities of daily living (ADL) in general is a popular topic in the field of human activity recognition and is often related to medical applications. Many of these applications utilize wearable sensors for ADL recognition, which is e.g. shown in [7]. A general work about the topic of human activity recognition with wearable sensors is presented in [42]. The techniques, which have been developed in this work, do not only treat the recognition of low-level activities (e.g. sitting, walking, standing, etc.), but also considers high-level activities (such as shopping, doing housework or commuting). High-level activities are modeled in here as a composition of sub-activities.

Surveillance Systems

Another area of research on human activity recognition are surveillance systems. In [6] human behavior in office environments is monitored with video cameras. The monitored behavior comprises the recognition of office work activities for an automated surveillance system. A more general approach for the recognition of office activities is presented in [72]. In this work, human behavior is modeled on different levels of abstraction. Within the European project Prometheus [4], a general framework for the interpretation of human behavior for no specific application has been developed. A different example for the usage of surveillance systems is the work of [77]. Here, an approach for the analysis of shopping behavior for product selling strategies is presented.

Consumer Electronics and Social Applications

Recognition of human activities plays also a role in future human computer interaction or assisting technologies, which are able to understand social human behavior [73]. The recognition of social activities is of interest for consumer electronics, such as mobile phones with integrated assisting technologies. An example for that is given by the work of [76], in which social behavior is modeled in form of interaction networks. A similar work is presented in [22], where mobile phones are utilized for analyzing social behavior. By the analysis of social behavior it is demonstrated that information like social relationships or daily user activity can be inferred from recorded mobile phone data. Another example for work about understanding social human behavior is given in [50], where the affect state of children is recognized while they perform a learning activity on the computer.

The above mentioned work gives a broad overview about the topic of human activity recognition and its applications. Basically, there has been much work published in this field of research of which components are transferable for application to particular industrial activity recognition problems. However, there is no system existing, which is able to satisfy our needs of a human-machine interface for the recognition of activities, gestures and tasks, which is flexible and appropriate for application in industrial environments.

2.2. Scenarios for Activity Recognition in Industrial Environments

Before developing an approach for a worker activity recognition system, which is able to serve as a human-machine interface for industrial environments, use cases have to be defined. This step is of particular importance, if the system should not only be designed for a specific application, but should consist of reusable components that may also be used in other scenarios. Based on our experience and the applications that can be found in literature, possible use cases have been divided into groups of tasks, which are performed by humans in industrial environments. The result is the following categorization of task-related use cases:

- Handling task: Common tasks for workers in industrial environments comprise the usage of tools, the operation of machines or handling of workpieces (e.g. material supply to machines). Characteristic for tasks of this group is that they contain a small number of simple steps, which are often repeated. Furthermore, process information can often be obtained from information about the internal machine state. In case of tool usage (such as welding guns, rivet guns, glue guns or nut runners), the tool pose provides useful clues for the status of the task, as well.
- Assembly or maintenance task: Assembly and maintenance tasks form another category of worker tasks. Tasks belonging to this category are usually of a higher complexity than handling tasks. During an assembly task, a product is assembled from workpieces in a logic sequential manner. Similar to this, maintenances are performed stepwise according to a maintenance plan. Because modern working environments are highly structured, it is characteristic for tasks of this category that they are executed as a sequence of predefined steps.
- Non-deterministic expert task: The category of non-deterministic expert tasks comprises tasks that are based on the occurrence of an unpredictable event and that are executed in a way, which depends on the occurred event. Examples for this are repair tasks (which deal with unidentified problems) or the supervision of processes. Since these tasks

are highly non-deterministic, the particular execution of the task is not known in advance. Therefore, tasks of this type are relatively complex and usually constitute intractable problems for task-related activity recognition.

The human-machine interface, which is presented in this work, aims at covering the recognition of tasks of the first two categories mentioned above. Because of the non-deterministic properties of tasks of the third class, activity recognition seems to be of minor use in here. However, this only refers to the (automatic) recognition of activities of the complexity of tasks and does not mean that the presented human-machine interface can not be of any help for use cases of this category. A helpful feature of a human-machine interface for non-deterministic expert tasks would be for instance the possibility of interaction with a superordinate framework in order to get work assistance or information.

Besides the recognition of performed tasks, the capability of comfortable interaction with a superordinate framework should generally be provided by a human-machine interface. Comfortable interaction means in this sense that a user can actively communicate (e.g. by issuing commands) via the interface without having to leave his workplace. An alternative to speech (which can be inconvenient in industrial environments because of noise) are gestures as a natural way of human communication. Thus, the recognition of human gestures is another use case to be considered.

For the demonstration of the task recognition results of this work, one representative scenario from each of the categories handling task and assembly or maintenance task have been chosen. Furthermore, comfortable interaction via hand gestures is additionally considered. The chosen scenarios are explained in the following.

It has to be noted that our activity recognition approach is demonstrated by means of these scenarios but is generally not limited to them. Because of the flexible solution concept, the system and its components can also be used in other scenarios of the treated task categories.

2.2.1. Spot Welding Scenario

A representative scenario for a handling task in the automotive industry is the spot welding scenario. In the scenario, a so-called hand welding gun is utilized, which is an industrial tool for manual resistance spot welding. The worker task in this scenario comprises activities such as positioning of the welding gun tip by alignment to a specific spot position or the execution of the welding process. The scenario finds application within the scope of XPRESS and is part of a demonstration where a human worker takes up the task of a welding robot after the occurrence of a malfunction in the robot.

2.2.2. Personal Computer Assembly Scenario

The personal computer (PC) assembly scenario is representative for worker tasks in the electrical industry of the category assembly or maintenance task. Since electronic devices are industrial products that are usually manufactured by manual assembly, the assembly of a PC constitutes a typical task of this category. The scenario consists of task steps, in which computer components (e.g. the mainboard or a drive) are built into the computer case. These task steps contain in turn activities, which are relevant for the execution of the task, such as picking up particular screws, placing and fastening a screw at the correct position, etc. By means of the PC assembly scenario a worker task of higher complexity than the spot welding task is treated. As part of a study in the XPRESS project, the scenario is exemplified in a laboratory test environment that is similar to a manual assembly workplace in the industry.

2.2.3. Hand Gesture Recognition

The recognition of hand gestures provides the possibility of comfortable interaction to the worker in order to get work assistance or task related information from a superordinate framework. In this scenario a limited number of predefined hand gestures are to be recognized. Because of the fact that interaction can also happen during the execution of a task, this scenario must not necessarily be treated in a stand-alone demonstration, but may also be integrated into one of the task recognition scenarios mentioned before.

2.3. Solution Concept

Based on the use cases and the scenarios of the previous section, a conceptual approach for the development of worker activity recognition systems has been defined. The solution concept utilizes a hierarchical model, which is depicted in figure 2.1. The hierarchical model describes the worker activity recognition process on four different levels of abstraction that range from sensor measurements up to reasoning about worker behavior of the complexity of tasks. By developing an activity recognition solution that is conform with this hierarchical model and by directing the development process in a top down manner, possible reutilization of components for new scenarios is facilitated.
For the description of the worker behavior, the following terminology is used:

Action

By actions, atomic worker behavior is described that is recognized from a source of processed sensor signals. A characteristic property of actions is that they are location invariant. Examples for actions are, for instance, a grasp that is recognized from a wearable sensor mounted to the hand of the worker or the execution of a weld, which is recognized from a welding gun trigger that has been pulled.

Activity

Similar to actions, activities (or low-level activities) represent atomic worker behavior, as well. However, activities are fused from detected actions and additional position information, and therefore, are not location invariant. The incorporation of location information in the activity recognition process is useful in particular, since in the structured environment of a factory, location information constitutes a valuable cue to currently performed activities. An example for an activity would be grasping a screw located in a particular box,

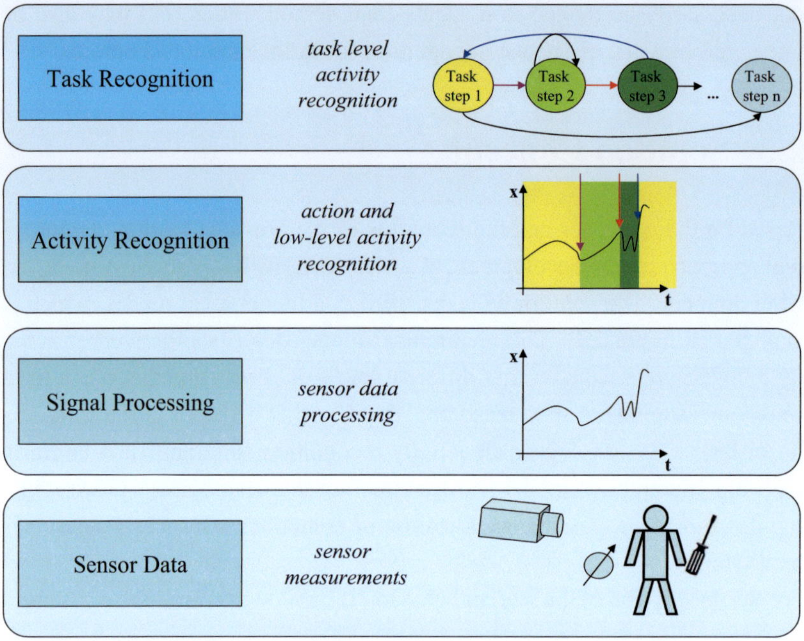

Figure 2.1: Hierarchical model describing the worker activity recognition process.

or in other words, a grasp action in combination with the worker hand being at the location of a screw box.

Task and subtasks

A task (or task level activity) describes worker behavior that is composed of several (low-level) activities or actions. Behavior that is recognized on the task level may also be a subtask, that is a certain part of a task. A worker task could be for instance the manual assembly of a PC and contain task steps, such as building in the mainboard.

The process levels of the hierarchical activity recognition model are explained in the following from the top to the bottom level.

2.3.1. Task Recognition Level

The task recognition level constitutes the highest level of abstraction. Since the execution of a worker task can generally be seen as a sequence of low-level activities or actions, activities and actions are treated as states or state transitions in a state-based task model. The first step in deriving the task model is to define the states by the decomposition of the task into a number of activities and actions, which are crucial for the completion of the task. The second step for deriving the task model is then to chose an adequate modelling method (e.g. a state machine) and to model the interrelations between the states. By means of modelling executed tasks as sequences of interrelated states we are able to make use of context knowledge [19] for task recognition. In the case if recognized activities or actions are uncertain (e.g. because they are affected by classification inaccuracies), a probabilistic task model would be able to recognize tasks from noisy inputs by making use of context information in form of interrelations between states.

2.3.2. Activity Recognition Level

On the activity recognition level, low-level activities are recognized by fusing information from recognized actions and location information that is derived from specific object positions. For action recognition, usually machine learning methods are utilized, which are trained to recognize actions from processed sensor data. In the design process, activities and actions that have to be recognized are defined by the needs of the task recognition level. However, for the choice of activity and action recognition methods the characteristics and availability of input information from the lower levels have to be considered.

2.3.3. Signal Processing Level

The signal processing level is an intermediate level for further processing of information from sensor measurements. Since raw data from sensors can often not directly be used as input information for activity recognition, signal processing is necessary to derive information, which is not directly available (e.g. object positions from images). Moreover, data from multiple sensor sources might be fused in order to increase the quality of obtained information [64]. Because of that, methods of the signal processing level and sensor devices often constitute combined systems and have to be chosen in dependence of each other in the design process.

2.3.4. Sensor Data Level

The sensor data level constitutes the link to the physical world. Besides the acquisition of sensor measurements, hardware design issues such as timestamp synchronization of different sensor sources have to be considered on this level. In the design phase of an interface, appropriate sensor devices are chosen for providing the information that is needed on higher levels of abstraction. However, it has to be considered that sensor device properties such as performance, price, size and availability limit the feasibility of the whole activity recognition system.

2.4. Human-Machine Interface Overview

An overview about the main components of the human-machine interface that has been developed in this work is depicted in figure 2.2. The human-machine interface has been developed according to the solution concept of the previous section and covers the scenarios from section 2.2. For having an overview about the human-machine interface and its components, a brief description is given in this section. The components of the interface and their design issues are explained in detail in the following chapters of the thesis.

For modelling and recognizing of task level activities from reliable activity recognition results a statechart model has been utilized. However, if activity recognition results are affected by uncertainty (e.g. possible classification errors) tasks are recognized by probabilistic methods. For probabilistic modelling of tasks, a hidden Markov model (HMM) has been used.

On the activity recognition level in figure 2.2, activities are derived by fus-

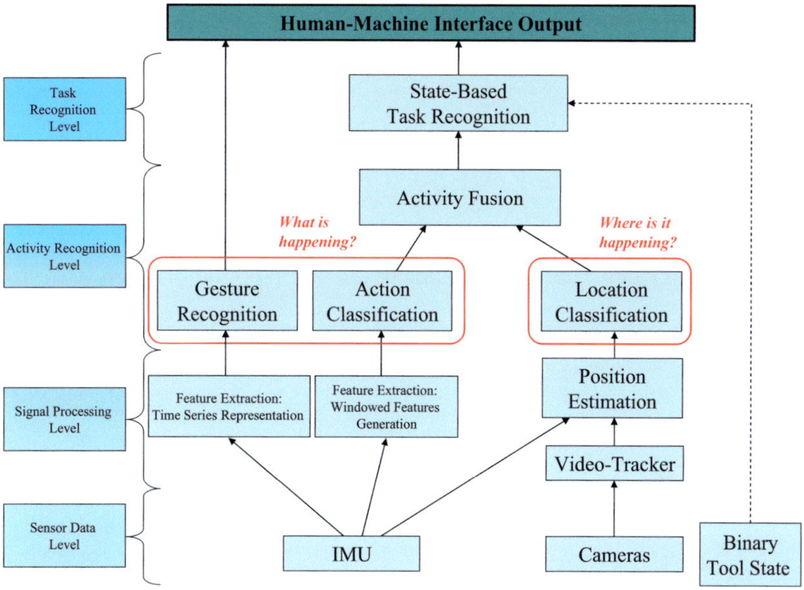

Figure 2.2: Overview about the main components of the human-machine interface.

ing location information (location classification) and actions detected from wearable sensor devices (action classification). Thus, the activity recognition approach does not only consider that an action is happening, but also takes into account where a particular action is happening. Similar to action classification, the location classification component utilizes machine learning for relating position areas to activities. Since the recognition of gestures is not involved in the task recognition process, gestures are treated as actions that are recognized separately. Because of the fact that gestures have significant temporal characteristics, a time series analysis method is utilized for gesture recognition.

On the signal processing level, features are extracted from sensor raw data for the gesture recognition component and the action classification component. Furthermore, position information is derived from a marker-based video-tracking system. Because of the fact that outages (e.g. caused by occlusions) can occur with vision-based position tracking, video-tracking results are fused with measurements from inertial sensors in order to derive better

position estimation results.
Utilized sensors on the sensor level are industrial cameras for position tracking such as inertial measurement units (IMU) for measuring the motion of specific objects (e.g. worker hands or tools) and for improving position tracking results. Optionally, the incorporation of additional binary state information of industrial tools (e.g. the welding gun trigger in the spot welding scenario) is considered. Since this type of sensor information contains no uncertainty it can directly be provided to the task recognition level.

3. Position Estimation System

All modern factories are characterized by structured working environments. Due to this fact, the execution of most of the activities is linked to particular locations. Therefore, information about the current position of moving objects, which are involved in the working process, contain a lot of valuable clues for reasoning about executed activities. These moving objects can be parts of the worker's body (e.g. hands), tools or even workpieces. Because of this reason, position estimation of objects in 3D space constitutes an important component in our activity recognition approach.

Depending on the application scenario, the recognition of human worker activities requires that a position estimation system is able to yield a certain accuracy, provide data at an adequate sampling rate and should be robust to disturbing environmental influences. This means that for being flexible with respect to the application scenario, positions should be measured in dimensions ranging from the order of meters for detecting the presence of humans at a workplace up to a few centimeters for recognizing the assembly of small pieces. For being able to capture quick motions such as hand movements, sampling rates of much more than 10 Hz should be provided. Concerning the robustness, it has to be regarded that an industrial environment can be rough with a lot of disturbances from machines or structures.

In this chapter, a system for indoor position estimation in industrial environments is presented. The system is able to provide position information of arbitrary objects by means of a marker-based video-tracking (VT) system. Depending on the setup, measurement accuracies in the order of centimeters or even millimeters at a sampling rate of several Hz can be achieved. Sampling rate and robustness to short outages are increased by means of an inertial navigation solution, which utilizes a small size MEMS (Micro-Electro-

Mechanical Systems) inertial measurement unit (IMU) and an extended Kalman filter.

The components of the approach, which is explained in the following sections, are depicted in figure 3.1. These components can also be found in the overview about the human-machine interface in figure 2.2. Parts of the presented position estimation approach can also be found in our publications [38] and [37].

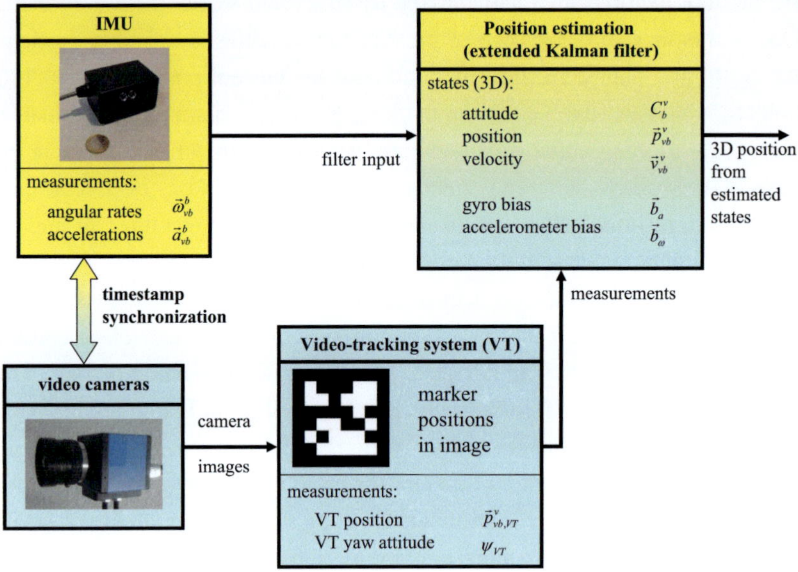

Figure 3.1: Components of the position estimation approach: IMU, video camera, VT system and position estimation with an extended Kalman filter.

3.1. Indoor Position Estimation Technology

In principle, indoor object positioning is possible with a variety of different sensors and a lot of different processing methods for estimating positions from sensor data are used for several types of applications. However, available techniques differ in accuracy, robustness, size and other important properties. Because of that, the choice of appropriate sensor devices and position estimation methods is crucial and should be done with the focus on the application. In order to give an overview about existing methods, commonly used sensing techniques for indoor positioning or pose estimation are listed in the following.

Range Measurement Techniques

Range measurement sensors measure distances from the sensor device to a measuring point by means of radio signals, optical signals or ultrasonic signals. Based on the measured distance, the position of the measuring point is usually determined either by incorporation of the direction, in which the sensor signal has been emitted, or by triangulation techniques (e.g. with multiple sensors of the same type). Depending on the application, sensor devices can be mounted on the object and detect known locations in the environment or are placed at known positions in the environment from which the object can be detected. A disadvantage of all range measurement techniques is that they are prone to errors from environmental influences such as occlusions, reflections or being out of range of the known or observed area.

A common technique for indoor position estimation by range measurements is the utilization of radio signals. Positioning systems based on radio signals estimate object positions by means of the travel time of radio signals (ToA, time of arrival). More accurate estimations can be achieved by so-called fingerprinting methods, which compare the current signal strength at the object to signal strengths at known locations. Among radio signal based techniques, Wi-Fi (Wireless Fidelity, synonymously used for Wireless Local Area Networks) systems are quite popular [84], due to the fact that more and more buildings are equipped with Wi-Fi access points. However, the positioning accuracy of these systems is limited and usually a resolution of not more than approximately a meter can be achieved [84].

A simple way to make use of radio signals is the utilization of the RFID (Radio Frequency Identification) technology. Utilization of defined objects [102] or proximity to specific locations can be detected by attaching an RFID reader to the object to be tracked and by marking the locations of interest with RFID tags. An inconvenience of this easy to use and cheap method is that location detection only works for a limited number of predefined positions.

The principle of distance calculations based on time of arrival measurements can also be applied to ultrasonic signals. Analogous to radio signal techniques, ultrasonic emitters and receivers are placed on an object and in the environment at known locations and in a specific arrangement in order to estimate the current object position. State of the art systems are able to yield an accuracy in the range of several centimeters [97, 111]. However, ultrasonic systems can suffer from reflection problems.

Besides the above mentioned techniques, other methods for indoor position estimation are existing, such as radar or laser scanners. Although the accuracy of these types of sensors can compete with others, their drawbacks are that they are expensive and comparably heavy. Therefore, they are rather suitable for mobile robot platforms than for human-machine interfaces.

Video Sensors

Similar to range measurement techniques, video-based positioning methods determine the position of an object by measuring the distance to fixed position references in the environment. However, since camera images only provide information of the appearance of an object in a scene, distances or position information have to be derived indirectly by means of image processing methods. Basically, two different approaches are existing for 3D position estimation with imaging sensors. The first one is to use a stereo or multi camera system and an object model for object identification. 3D positioning can then be done by optical triangulation. An alternative are marker-based VT systems, which utilize fiducial markers that have to be attached to an object or be placed at known positions in the environment. The markers facilitate object identification and some allow 3D position estimation with a single camera.

Vision-based systems are very common for indoor position estimation, especially in applications related to the analysis or tracking of human movements [3, 32, 66, 113]. A very useful advantage of vision-based object tracking systems is that their accuracy depends on the measurement setup (e.g. resolution of the camera and object size). Therefore, they are flexible in usage and can be used for object tracking in spaces with small dimensions, but also for large scale applications. However, a drawback of vision-based position estimation is that complex image processing methods with comparable high computing costs are required. Another disadvantage of camera sensors is that line of sight and illumination problems may occur.

Inertial Sensors

Inertial sensors measure motion related physical quantities in form of accelerations and angular rates, which occur directly at the sensor device. Assuming that initial position and attitude are known, the current sensor position can be derived from these measurements by means of dead reckoning methods. Therefore, inertial sensors are mounted on a moving object and theoretically need no specific setup or location references in the environment for position estimation. The development of small and affordable inertial sensors based on MEMS technology allows it to utilize inertial sensors in a broad field of applications.

In contrast to remote sensing techniques, inertial sensing techniques are always able to provide position estimations, because they are not influenced by outage producing environmental disturbances. However, a major drawback of inertial sensing techniques are drift problems, which are caused by biased sensor signals and cumulated dead reckoning errors. Therefore, position estimations based on inertial sensors alone can only be accurate for short periods of time and need to be aided by other sensing techniques.

Long-term drift free attitude estimations are possible by incorporating acceleration measurements as vertical reference [83]. Besides the usage of aiding sensor techniques for position estimations, drift problems can be mitigated by incorporation of movement models in filter methods, which indicate particular movement states, such as zero velocity updates for pedestrian navigation [71] [10]. Alternatively, the pose of the human body (or of other joint like objects) can be estimated by utilizing attitude estimations and a model of limbs and joints [128].

Multisensor Techniques

For some applications, position estimations based on a single type of sensor device are not able to meet the requirements of the desired system. Therefore, it makes sense to take advantage of the complimentary properties of several types of sensors and to gain better estimates by sensor fusion [64]. This principle is widely used in inertial navigation systems (INS) [105]. An example

on the large scale is that INS based on measurements from high-grade IMUs are often aided by measurements from a global positioning system (GPS) for aircraft navigation.
On the small scale, different indoor applications have been developed in recent years, which utilize inertial sensors (such as MEMS IMUs) together with other sensing technologies for position estimation. A lot of applications favor the combination of inertial sensors and vision-based methods [16] for compensation of inertial sensor drift with long term accurate measurements from video sensors. Applications cover fields of research like augmented reality [91, 125], navigation for micro flyers [89], medical engineering [104] or industrial tool tracking [75]. While inertial sensors always have to be mounted on the object to be tracked, cameras do not necessarily have to be attached to the object as well. Placing cameras at locations in the environment for tracking the object and aiding the inertial navigation solution with position estimates [74] works analogous to the GPS-based solution. The alternative is to attach the camera to the object in order to observe the environment for landmarks or location cues [29, 41]. [28] even incorporates both methods in a tracking solution for achieving higher positioning accuracies.
Besides position estimation systems based on inertial sensors and vision-based methods, other multi-sensor solutions are existing, such as the combination of inertial sensors with ultrasonic sensors [111].

For our application of estimating object positions in industrial environments, we decided for a system based on a marker-based VT system and inertial sensors. This combination of sensors has complementary benefits and appears to be most appropriate for our application.
Important advantages of marker-based VT are that this solution is scalable and, compared with other vision-based methods, independent of the application, since no object depending model is required. Furthermore, video sensors have no problems with common influences of industrial environments like reflections from metal structures, electromagnetic fields from electric drives, noise, etc. Regarding the requirement that the system should be able to estimate positions of arbitrary objects at different position ranges, this technique seemed to be the best choice. For dealing with outages and limited

sampling rates of the VT system, VT measurements are fused with data from a MEMS IMU by means of an extended Kalman filter.

3.2. Sensor Hardware and Video-Tracking System

This section explains the sensor hardware components and the VT system component of the position estimation system in figure 3.1. The sensor hardware consists of MEMS IMUs mounted on connector boards, providing a PC interface and hardware synchronization, such as triggerable video cameras.

3.2.1. Inertial Sensors and Hardware Synchronization

For measuring accelerations and angular rates of objects in 3D space, inertial measurement units are utilized, which are mounted on each object to be tracked. The utilized IMUs (type ADIS16350 from Analog Devices [107]) contain 3 orthogonal accelerometers and 3 orthogonal gyroscopes and are configured to provide measurements at a sampling rate of 407 Hz. This sampling rate is high enough to capture even very fast motions of human body parts or used tools. The characteristic noise of the sampled signals has standard deviations of $0.4°/s$ for gyroscopes and $0.05m/s^2$ for accelerometers. Calibration of the IMU without professional equipment can be done by a procedure described in [25]. However, calibration of IMU biases, which are the main source of error in EKF estimations, is unhelpful, since biases are only stable for short periods of time. Thus, they can only be determined by estimation in the EKF model (as explained in section 3.3).

Hardware Synchronization

In order to provide measurements with correctly assigned timestamps, utilized sensor devices have to be hardware synchronized. Therefore, all sensors are operated in a sensor network containing microcontroller connector boards, on which the IMUs are mounted. Each connector board assigns timestamps to sampled IMU measurements and transmits the measured data to a PC via an UART interface. As depicted in figure 3.2, the hardware synchronization in the sensor network is realized by a master board, which synchronizes microcontroller timers of other connector boards, if more than one object is tracked. Furthermore, camera sensors are triggered by the master board. This makes it possible that timestamps are taken when camera images are triggered. Image timestamps are sent from the master board to the PC where they can be assigned to arriving images.

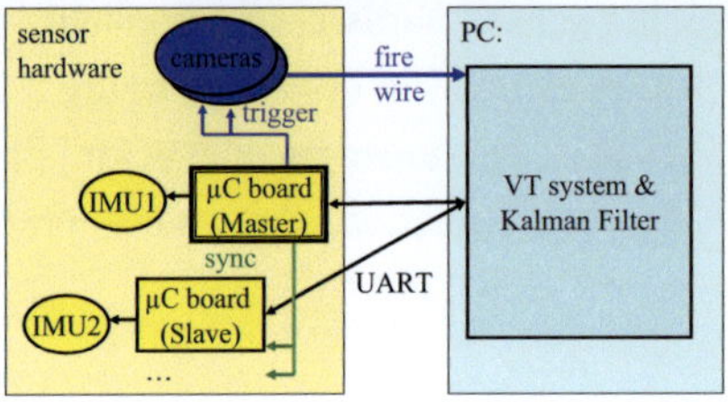

Figure 3.2: Scheme for hardware synchronization of IMUs and cameras. Synchronization is done by means of microcontroller (μC) connector boards. Camera images are triggered and μC clocks are synchronized by means of a master controller.

3.2.2. Video Cameras and Camera Calibration

The marker-based VT system processes the images of video cameras, which are placed in the environment for the detection of VT markers in the workplace region. In our system, triggerable industrial cameras (type DBK21BF04 from the company The Imaging Source [109]) are used, which are connected to a PC via firewire (IEEE1394). In order to have a broad field of view, the cameras have been equipped with wide-angle lenses.

As explained in the previous section, timestamps of camera images are acquired by means of hardware synchronization. Timestamp inaccuracies caused by latencies introduced by the exposure time are kept small by choosing short exposure times (typically $1/1000s$). Because of that, it has to be taken care in measurement setups that the workplace is well-illuminated.

Camera Calibration

Before being able to use a camera with the VT system, the intrinsic parameters of the camera model have to be determined. The camera model describes the projection of points in the camera coordinate system $\vec{p}^c = (x,y,z,1)^T$ [1] to pixel coordinates of the image plane $\vec{p}^i = (u,v,1)^T$. Parameters of this model are the intrinsic camera parameters, which comprise lens distortion coefficients, the focal length f, the principal point (x_c, y_c), the image format (s_x, s_y) and the skew coefficient γ.

Lens distortion occurs especially with wide-angle lenses, that is lenses with a small focal length. Therefore, the lens distortion model has to be considered for avoiding the introduction of additional errors in the tracking result by the utilized wide-angle lenses. Lens distortions are described by a parametric model, which can be found in [13]. The model incorporates radial and tangential lens distortion by means of the following equation, which relates the camera coordinates $\vec{p}^c$ to the distorted projection point $\vec{p}^d = (x_d, y_d, 1)^T$:

$$\begin{aligned} \begin{pmatrix} x_d \\ y_d \end{pmatrix} &= \begin{pmatrix} x + \tilde{x}\left(k_1 r^2 + k_2 r^4 + k_3 r^6 + \ldots\right) \\ y + \tilde{y}\left(k_1 r^2 + k_2 r^4 + k_3 r^6 + \ldots\right) \end{pmatrix} \\ &+ \begin{pmatrix} \left(p_1\left(r^2 + 2\tilde{x}^2\right) + 2p_2\tilde{x}\tilde{y}\right)\left(1 + p_3 r^2 + \ldots\right) \\ \left(p_2\left(r^2 + 2\tilde{y}^2\right) + 2p_1\tilde{x}\tilde{y}\right)\left(1 + p_3 r^2 + \ldots\right) \end{pmatrix}, \end{aligned} \tag{3.1}$$

where $\tilde{x} = x - x_c$, $\tilde{y} = y - y_c$ and $r = \sqrt{\tilde{x}^2 + \tilde{y}^2}$. The distortion coefficients $(k_1, k_2, \ldots)$ and $(p_1, p_2, \ldots)$ in equation (3.1) are to be found by a camera calibration routine and describe the radial and tangential lens distortion.

The other intrinsic parameters of the camera model relate the distorted projection point $\vec{p}^d$ to pixel coordinates $\vec{p}^i$ by the camera matrix:

$$\vec{p}^i = K\vec{p}^d = \begin{pmatrix} \alpha & \gamma & x_c \\ 0 & \beta & y_c \\ 0 & 0 & 1 \end{pmatrix} \vec{p}^d. \tag{3.2}$$

By means of the coefficients α and β in equation (3.2), the focal length is related to pixels in each image direction according to $\alpha = s_x f$ and $\beta = s_y f$.

Different tools and libraries are available for camera calibration. A widely used tool is the Camera Calibration Toolbox for Matlab [108], which partially bases on the work of Zhang [127]. The video cameras that are utilized in this work are calibrated with this calibration software.

[1] The notation utilized in here is called homogeneous coordinates. Homogeneous coordinates allow the expression of a coordinate system transformation of a position vector by means of only one matrix multiplication.

3.2.3. Video-Tracking System

The utilized VT system is based on a modified version of the ARToolKitPlus [112] library. ARToolKitPlus is an extension of the augmented reality library ARToolKit [52]. By means of the ARToolKitPlus library, 6 degrees of freedom (DOF), that is position and orientation, of a fiducial marker can be measured in 3D space with a single camera. Fiducial markers are 2-dimensional square patterns of known size, which are attached to objects to be tracked. An example for a marker can be seen in the box of the VT component in figure 3.1. The marker pattern consists of a 6×6 BCH code, which allows robust detection of 4096 different rotation invariant markers. Thus, several markers can be used at the same time in a redundant way to increase the detection robustness of a single object and for tracking multiple objects.
The eventual objective of the VT system is the estimation of object position $\vec{p}^{v}_{vb,VT} = (x_{VT}, y_{VT}, z_{VT})^T$ and attitude $\vec{\Psi}_{VT} = (\phi, \theta, \psi)^T$ (here given in Euler angles - see also section 3.3.1 for the Euler angle representation) in VT coordinates. In our application, the VT coordinate system coincides with the world coordinate system, which constitutes the workplace environment. The functionality of ARToolKit allows the estimation of the extrinsic camera parameters, that is to determine the camera position and orientation in the marker coordinate system. Thus, a point $\vec{p}^b$ given in an object coordinate system can be related to a point in camera coordinates $\vec{p}^c$ by means of the following equation:

$$\vec{p}^c = T_b^c \vec{p}^b = \begin{pmatrix} R_b^c & \vec{t_{cb}} \\ (0\,0\,0) & 1 \end{pmatrix} \vec{p}^b, \tag{3.3}$$

in which R_b^c and $\vec{t_{cb}}$ constitute the rotational and translational components of the relation between camera and object.
The relation of the object to be tracked to the video coordinate system can now be determined with the transformation matrix T_v^c according to:

$$T_b^v = (T_v^c)^{-1} T_b^c. \tag{3.4}$$

The object position $\vec{p}^{v}_{vb,VT}$ and attitude $\vec{\Psi}_{VT}$ in VT coordinates are then given by the rotational and translational components of T_b^v.
Equation (3.4) bases on the knowledge of the relation between camera coordinates and VT coordinates T_v^c. Therefore, the transformation matrix T_v^c has

to be determined by means of a calibration procedure, which has to be applied once for each new measurement setup, in which the VT system is used. The VT system setup calibration works in the same way as the extrinsic camera parameters are measured. T_v^c can simply be determined, if a marker object is placed in the origin of the VT coordinate system.

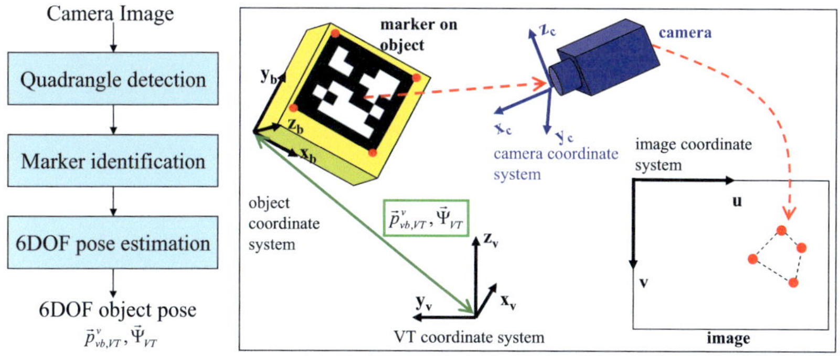

Figure 3.3: Left: Flow chart of the ARToolKit pose estimation algorithm. Right: Illustration of the pose estimation process. A marker in the camera field of view is projected on the image plane. With the knowledge of the intrinsic camera parameters and the marker geometry, the extrinsic camera parameters are determined by the ARToolKit algorithm. From the extrinsic camera parameters and the known transformation of camera coordinates to video coordinates, the object pose in the Video coordinate system is determined.

On the left of figure 3.3, a flow chart of the most important processing steps of the ARToolKit pose estimation process is shown. In the first step, quadrangles are detected from intersecting edges in the camera image. Each detected quadrangle is treated as a marker candidate, which has to be identified in the second step. In the second step, the quadrangles are rectified and scaled to patterns of 6×6 pixels. After binarization of the extracted patterns, it can be checked, whether a pattern corresponds to a marker ID or not. In the 6DOF pose estimation step, the extrinsic camera parameters are determined. An illustration of mappings in the pose estimation process can be found in the box on the right of figure 3.3. The extrinsic camera parameters R_b^c and $\vec{t_{cb}}$ are determined from the projected points on the image plane, which correspond to known corner point positions of the marker in the object coordinate system. Determining the 6DOF object pose is possible with a single camera by incor-

poration of knowledge about the intrinsic camera parameters and the marker size.
The modified version of ARToolKitPlus utilized in the VT system provides better illumination robustness and allows subpixel-accurate marker detection. Another feature of this modified version is that multiple cameras can be utilized for increasing the field of view and improving the pose estimation result.

3.3. Inertial Navigation Solution for Position Tracking

The concept of the position estimation system is based on measurements from two complementary types of sensor systems, an IMU and a VT system. Estimating fused positions and attitudes with this sensor combination can be realized by means of an inertial navigation system solution [34, 105]. The calculation of position and attitude from IMU measurements is achieved in the INS by integration with a so-called strapdown algorithm. Due to the fact that IMU measurements are affected by errors, such as biases and noise, integrated errors lead to a drift, which increases with time. The drift can be mitigated by fusing results of the strapdown algorithm with measurements of the VT system. As a result, fused position and attitude estimations combine the advantages of accurate and scalable pose measurements of the VT system and an increased sampling rate such as compensation of short outages by means of integrated IMU measurements.
For solving an inertial navigation problem, different types of stochastic filters may be utilized [116]. One of the most commonly used filter types is the Kalman filter (KF) [49]. Because of the fact that the KF is designed to solve linear problems, the extended Kalman filter (EKF) constitutes a better choice for nonlinear problems. This must not be the case, if the INS design is based on an error-state model. The error-state model approach estimates navigation errors with a linearized model, which is convenient for embedded systems on small platforms with low computing power. However, in [45] it has been shown that a total state space filter (which can be realized by using the EKF) is able to yield a better estimation performance. The unscented KF

or the particle filter are filter alternatives suitable for highly nonlinear filtering problems. However, a drawback is that they consume much more computing time than the EKF [116]. Since the VT system already consumes much computing time, utilization of an unscented KF or a particle filter would reverse one of the desired benefits of the INS solution. Therefore, an EKF is used for the position estimation system.

3.3.1. Attitude Representations and Strapdown Model

For the mathematical description of a navigation system, different coordinate systems are needed [105]. The presented position estimation system utilizes two different coordinate systems. IMU measurements are taken in the object coordinate system or b-frame and are denoted with $()^b$. The b-frame is fixed to the tracked object. VT measurements are taken in the VT coordinate system or v-frame, which is fixed to the workplace environment. State variables or measurements in the VT coordinate system are denoted with $()^v$.

Attitude Representations

The transformation of a vector from one coordinate system representation into another coordinate system representation contains a rotational and a translational component. However, the translational component can be omitted, either if the coordinate systems have the same origins or if the transformed vector is merely a direction vector and no position vector. The latter case is for attitude representations in INS. This fact allows it to express coordinate transformations by means of a 3×3 rotation matrix, also called direction cosine matrix. For instance, accelerations given in the b-frame are transformed into the v-frame by:

$$\vec{a}_{vb}^{v} = C_b^v \vec{a}_{vb}^{b}. \tag{3.5}$$

A more comprehensible attitude representation is given by the Euler angles $\vec{\Psi} = (\phi, \theta, \psi)^T$. Euler angles represent rotations about the coordinate system axes, which are performed after another. In the convention used for INS, the first rotation is around the z-axis by yaw ψ. The second and third rotations are around the new y-axis by pitch θ and around the resulting x-axis by roll ϕ, respectively. The rotation matrix is expressed by Euler angles in the following way:

$$C_b^v = \begin{pmatrix} c\psi c\theta & c\psi s\theta s\phi - s\psi c\phi & c\psi s\theta c\phi + s\psi s\phi \\ s\psi c\theta & s\psi s\theta s\phi + c\psi c\phi & s\psi s\theta c\phi - c\psi s\phi \\ -s\theta & c\theta s\phi & c\theta c\phi \end{pmatrix}. \tag{3.6}$$

In equation (3.6), the trigonometric functions $sin()$ and $cos()$ are abbreviated with s and c.

Attitude changes resulting from angular rates $\vec{\omega}_{vb}^b$ are described by means of the following differential equations:

$$\dot{\phi} = \left(\omega_{vb,y}^b sin\phi + \omega_{vb,z}^b cos\phi\right) tan\theta + \omega_{vb,x}^b, \tag{3.7}$$

$$\dot{\theta} = \omega_{vb,y}^b cos\phi - \omega_{vb,z}^b sin\phi, \tag{3.8}$$

$$\dot{\psi} = \left(\omega_{vb,y}^b sin\phi + \omega_{vb,z}^b cos\phi\right) sec\theta. \tag{3.9}$$

A disadvantage of Euler angles is a singularity, which occurs at $\theta = \pm 90°$. Solutions of equations (3.7) and (3.9) are not defined at this singularity.

Another way to express rotations is the utilization of a rotation vector $\vec{\sigma} = (\sigma_x, \sigma_y, \sigma_z)^T$. The components of the rotation vector constitute a rotation axis in space about which a coordinate system is rotated. The length of the rotation vector $|\vec{\sigma}|$ specifies the magnitude of the rotation. Rotation axis and magnitude of the rotation can be used as components of a normalized quaternion:

$$\vec{q} = \begin{pmatrix} a \\ b \\ c \\ d \end{pmatrix} = \begin{pmatrix} cos\left(|\vec{\sigma}|/2\right) \\ \left(\sigma_x/|\vec{\sigma}|\right) sin\left(|\vec{\sigma}|/2\right) \\ \left(\sigma_y/|\vec{\sigma}|\right) sin\left(|\vec{\sigma}|/2\right) \\ \left(\sigma_z/|\vec{\sigma}|\right) sin\left(|\vec{\sigma}|/2\right) \end{pmatrix}. \tag{3.10}$$

The propagation of attitude given in the quaternion representation is expressed by the differential equation

$$\dot{\vec{q}}_b^v = \frac{1}{2}\vec{q}_b^v \bullet \begin{pmatrix} 0 \\ \vec{\omega}_{vb}^b \end{pmatrix}. \tag{3.11}$$

The change of the attitude $\vec{q}_b^v$ in equation (3.11) is proportional to the quaternion product of the attitude quaternion and measured angular rates. For a more detailed explanation of the quaternion product operator $\bullet$ and quaternion mathematics see also [105] or [116].

An advantage of the quaternion notation over Euler angles is that there is no singularity, at which the update differential equation becomes indeterminable. The rotation matrix C_b^v can be expressed by the quaternion elements as:

$$C_b^v = \begin{pmatrix} a^2+b^2-c^2-d^2 & 2(bc-ad) & 2(bd+ac) \\ 2(bc+ad) & a^2-b^2+c^2-d^2 & 2(cd-ab) \\ 2(bd-ac) & 2(cd+ab) & a^2-b^2-c^2+d^2 \end{pmatrix}. \quad (3.12)$$

Strapdown Equations

By means of the strapdown algorithm, the current position, velocity and attitude are calculated from their initial values and IMU measurements, which are measured over a certain period of time. A block diagram of the strapdown model is depicted in figure 3.4.

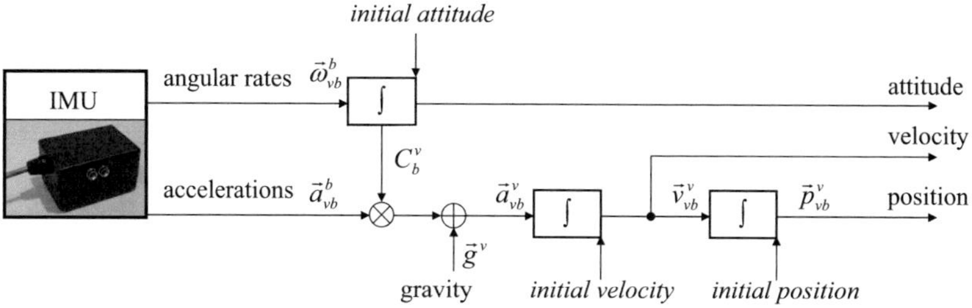

Figure 3.4: Block diagram of the strapdown model.

Assuming that the attitude C_b^v is known, accelerations measured in the b-frame are transformed to the v-frame in the strapdown model. After this transformation, we are able to correct the influence of the local gravity $\vec{g}^v$ in measured accelerations. The resulting acceleration $\vec{a}_{vb}^v$ then constitutes the change of velocity relative to the v-frame. The following differential equation describes the dynamics of the object velocity:

$$\dot{\vec{v}}_{vb}^v = \vec{a}_{vb}^v + \vec{g}^v = C_b^v \vec{a}_{vb}^b + \vec{g}^v. \quad (3.13)$$

Deriving the current position is then straightforward:

$$\dot{\vec{p}}_{vb}^v = \vec{v}_{vb}^v. \quad (3.14)$$

It has to be noted that the utilized inertial sensors in the MEMS IMU are not sensitive enough to measure the turn rate of the earth. Because of that and because of the fact that a workplace has limited dimensions, turn rate of the earth, such as transport rate, can be neglected. Furthermore, since comparable small velocities occur with human motion, Coriolis effects can be neglected as well.

The equations utilized in the strapdown algorithm are the discrete solutions to the differential equations, which describe the continuous strapdown model. For the propagation of estimated quantities in time, it is assumed that IMU measurements remain (nearly) constant over the sampling intervals ΔT.

The differential equations describing the propagation of Euler angles given in equations (3.7)-(3.9) can be expressed by the following discrete solution:

$$\vec{\Psi}_{k+1} = \vec{\Psi}_k + \begin{pmatrix} \left(\omega^b_{vb,y} s\phi + \omega^b_{vb,z} c\phi\right) tan\theta + \omega^b_{vb,x} \\ \omega^b_{vb,y} c\phi - \omega^b_{vb,z} s\phi \\ \left(\omega^b_{vb,y} s\phi + \omega^b_{vb,z} c\phi\right) sec\theta \end{pmatrix}_k \Delta T. \tag{3.15}$$

If the quaternion attitude representation is utilized, the new attitude $\vec{q}^{\,v}_{b,k+1}$ is calculated by a quaternion multiplication of the old attitude $\vec{q}^{\,v}_{b,k}$ with an update quaternion $\vec{r}^{\,b}_{b}$, which is the solution to the quaternion differential equation (3.11):

$$\vec{q}^{\,v}_{b,k+1} = \vec{q}^{\,v}_{b,k} \bullet \vec{r}^{\,b}_{b}. \tag{3.16}$$

The update quaternion in equation (3.16) can be derived from equation (3.10) and the rotation vector:

$$\vec{\sigma} = \vec{\omega}^b_{vb} \Delta T. \tag{3.17}$$

Velocities are propagated in time by the solution of equation (3.13):

$$\vec{v}^{\,v}_{vb,k+1} = \vec{v}^{\,v}_{vb,k} + \Delta T \left(C^v_b \vec{a}^{\,b}_{vb,k} + \vec{g}^{\,v}\right). \tag{3.18}$$

Finally, the following equation contains the discrete solution of the position differential equation (3.14):

$$\vec{p}^{\,v}_{vb,k+1} = \vec{p}^{\,v}_{vb,k} + \vec{v}^{\,v}_{vb,k} \Delta T + \frac{1}{2} \Delta T^2 \left(C^v_b \vec{a}^{\,b}_{vb,k} + \vec{g}^{\,v}\right). \tag{3.19}$$

3.3.2. Kalman Filter Models

For estimating positions and attitudes from IMU measurements and VT results with the KF component in figure 3.1, a total state space KF is used. The utilized EKF model estimates the 9-dimensional state vector

$$\vec{x}_k = \left(\vec{\Psi}_k{}^T,\; \vec{p}^{\,v}_{vb,k}{}^T,\; \vec{v}^{\,v}_{vb,k}{}^T \right)^T, \tag{3.20}$$

which consists of the following states:

- attitude in Euler angles $\vec{\Psi} = (\phi, \theta, \psi)^T$,
- position $\vec{p}^{\,v}_{vb} = (x, y, z)^T$,
- velocity $\vec{v}^{\,v}_{vb} = (v_x, v_y, v_z)^T$.

Because of the fact that biases in IMU measurements are not stable, they cannot adequately be removed by calibration. Therefore, the sensor biases are incorporated in an alternative system model based on 15 states and resulting in the state vector:

$$\vec{x}_k = \left(\vec{\Psi}_k{}^T,\; \vec{p}^{\,v}_{vb,k}{}^T,\; \vec{v}^{\,v}_{vb,k}{}^T,\; \vec{b}_{\omega,k}{}^T,\; \vec{b}_{a,k}{}^T \right)^T. \tag{3.21}$$

The state vector in equation (3.21) contains the following bias states:

- angular rate sensor bias $\vec{b}_\omega = (b_{\omega x}, b_{\omega y}, b_{\omega z})^T$,
- accelerometer bias $\vec{b}_a = (b_{ax}, b_{ay}, b_{az})^T$.

The state vectors in equations (3.20) and (3.21) constitute the basis of two different filter models. Both EKF models exist also for the quaternion attitude representation. In this case, Euler angles $\vec{\Psi}$ are substituted by the quaternion $\vec{q}^{\,v}_b = (a, b, c, d)^T$ in the state vector, which increases its dimension to 10 or 16 states, respectively.

The propagation of the state vector in each model is described by a nonlinear system model function $\vec{f}(\vec{x}_k, \vec{u}_k, \vec{w}_k)$ according to:

$$\vec{x}_{k+1} = \vec{f}(\vec{x}_k, \vec{u}_k, \vec{w}_k). \tag{3.22}$$

Equation (3.22) contains the strapdown equations (3.19), (3.18) and (3.15) (which is substituted with equation (3.16) if the quaternion representation is used).

For the alternative filter model, additional equations are used in the system model function in order to describe the propagation of biases:

$$\vec{b}_{\omega,k+1} = \vec{b}_{\omega,k}, \tag{3.23}$$

$$\vec{b}_{a,k+1} = \vec{b}_{a,k}. \tag{3.24}$$

Since the IMU biases change rather slowly with time, they are assumed to be constant in equations (3.23) and (3.24).
The input vector $\vec{u}_k$ consists of the accelerometer readings $\vec{a}^b_{vb,IMU,k}$ and the angular rate sensor readings $\vec{\omega}^b_{vb,IMU,k}$ of the IMU:

$$\vec{u}_k = \left(\vec{\omega}^b_{vb,IMU,k}{}^T, \; \vec{a}^b_{vb,IMU,k}{}^T \right)^T. \tag{3.25}$$

Accelerometers and angular rate sensors are modeled by means of sensor models, which are given in the following equations:

$$\vec{\omega}^b_{vb,IMU} = \vec{\omega}^b_{vb} + \vec{b}_\omega + \vec{w}_\omega, \tag{3.26}$$

$$\vec{a}^b_{vb,IMU} = \vec{a}^b_{vb} + \vec{b}_a + \vec{w}_a. \tag{3.27}$$

In equations (3.26) and (3.27), the dominant error characteristics of the IMU sensors are described as biases $\vec{b}_\omega$, $\vec{b}_a$ and noise $\vec{w}_\omega$, $\vec{w}_a$. The utilized sensor models are a reduced form of precise sensor models of high quality IMUs, which can be found in [105]. The sensor noise $\vec{w}$ is assumed to be normally distributed and zero mean.
Measurement updates from the VT system are considered by means of the measurement model

$$\vec{z}_k = \vec{h}\left(\vec{x}_k, \vec{v}_k\right). \tag{3.28}$$

For compensation of the drift of position and velocity estimates in the strap-down model, position updates have to be utilized. Furthermore, the yaw angle is not observable with the described EKF models, if the tracked object is not moving (see also [116]). Therefore, it should be considered for updates, as well. Incorporation of the other attitude measurements from the VT system is only advisable if multiple markers are used for VT attitude estimations. Otherwise, attitude estimations with the VT system are usually not very accurate. Thus, the measurement vector $\vec{z}_k$ contains VT position and yaw angle measurements:

$$\vec{z}_k = \left(\psi_{VT}, \; \vec{p}^v_{vb,VT}{}^T \right). \tag{3.29}$$

The measurement function $\vec{h}\left(\vec{x}, \vec{v}\right)$ then consists of the estimated yaw angle and the estimated position vector, such as the normally distributed measurement noise $\vec{v}$:

$$\vec{h}\left(\vec{x}, \vec{v}\right) = \begin{pmatrix} \psi \\ \vec{p}^v_{vb} \end{pmatrix} + \vec{v}. \tag{3.30}$$

If a quaternion attitude representation is used, ψ has to be substituted in equation (3.30) by the respective quaternion expression.
With the system model function in equation (3.22) and the measurement function in equation (3.30) we are now able to determine the filter matrices of the EKF as the Jacobian matrices given in the following equations:

$$\Phi_k = \frac{\partial \vec{f}}{\partial \vec{x}}\left(\vec{x}_k, \vec{u}_k, 0\right), \tag{3.31}$$

$$W_k = \frac{\partial \vec{f}}{\partial \vec{w}}\left(\vec{x}_k, \vec{u}_k, 0\right), \tag{3.32}$$

$$H_k = \frac{\partial \vec{h}}{\partial \vec{x}}\left(\vec{x}_k, 0\right), \tag{3.33}$$

$$V_k = \frac{\partial \vec{h}}{\partial \vec{v}}\left(\vec{x}_k, 0\right). \tag{3.34}$$

The EKF algorithm consists of a prediction step and correction step. In the prediction step, the system state vector is propagated by means of the input information from IMU measurements $\vec{u}_k$:

$$\vec{x}_{k+1}^{-} = \vec{f}\left(\vec{x}_k^{+}, \vec{u}_k, 0\right). \tag{3.35}$$

The system covariance is propagated in the prediction step based on the covariance matrix of the IMU measurement noise Q_k:

$$P_{k+1}^{-} = \Phi_k P_k^{+} \Phi_k^T + W_k Q_k W_k^T. \tag{3.36}$$

The correction step is executed, when new VT system measurements $\vec{z}_k$ are available. System state and system covariance are updated with these measurements according to the following equations:

$$K_k = P_k^{-} H_k^T \left(H_k P_k^{-} H_k^T + V_k R_k V_k^T\right)^{-1}, \tag{3.37}$$

$$\vec{x}_k^{+} = \vec{x}_k^{-} + K_k \left(\vec{z}_k - \vec{h}\left(\vec{x}_k^{-}, 0\right)\right), \tag{3.38}$$

$$P_k^{+} = \left(I - K_k H_k\right) P_k^{-}, \tag{3.39}$$

in which R_k is the covariance matrix of the measurement noise.
Q_k and R_k may also be used as filter tuning parameters.

3.4. Experiments and Results

In this section, results from experiments with the proposed position estimation system are shown. The experiments have been conducted for evaluating the performance of the inertial navigation solution with a simulated dataset and with data from a realistic scenario. Additionally, an account of the positioning accuracy of the VT system is given for the realistic scenario.

3.4.1. Experiments with Simulated Data

For evaluation of the INS solution under ideal conditions, tests have been conducted with a set of simulated trajectories. The advantage of tests with simulated data is that erroneous influences on measurements, such as noise, biases, synchronization errors or outages, can be evaluated depending on their specification. Furthermore, the simulated dataset contains the ground truth of all estimated states, which can be used as an exactly known reference. In all tests, EKF models based on the Euler angle representation have been used for obtaining a comprehensible account of attitude estimation results.

Simulated Dataset

Simulated motion data has been created by defining a course of accelerations and angular rates of a fictive movement in 3D space of 40s duration. The corresponding trajectories of velocity, position and attitude have been derived from the motion data by application of a strapdown algorithm. The calculated trajectories constitute the ground truth data, which has been used as a reference for EKF estimation results.

Simulated IMU input data has been generated by adding normally distributed noise to simulated accelerations and angular rates. Furthermore, constant offset values have been added to the input data in order to simulate biased IMU signals. Update measurements of the VT system have been simulated by sampling the ground truth data at a reduced and constant sampling rate. After sampling, normally distributed measurement noise has been added to simulated VT data.

All sampling rates, biases and noise characteristics have been chosen in a

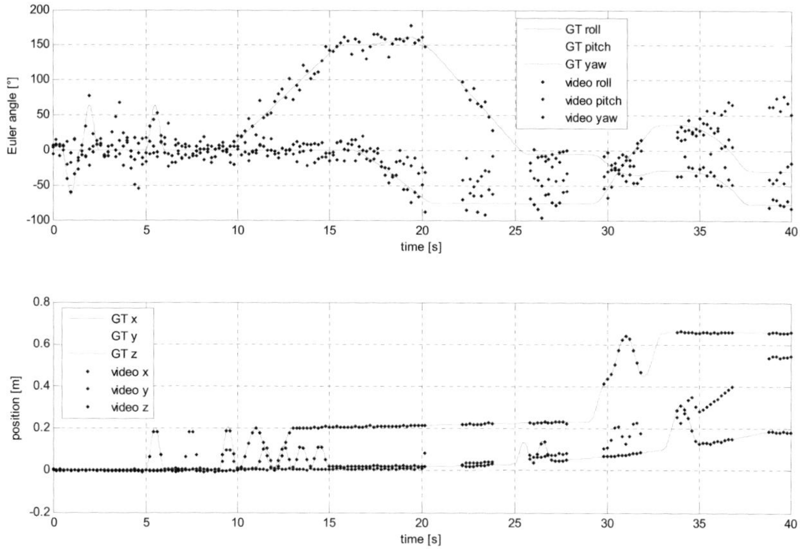

Figure 3.5: VT measurements (attitudes and positions) of a simulated test run with outages. Ground truth data is represented by lines and sampling points of simulated VT measurements by dots.

range of values, which is typical for the hardware in section 3.2. By means of this procedure, a set of simulated measurements has been created in 10 runs with the same underlying movement, but randomly varying superimposed noise. Furthermore, the set has been extended by 10 series that have been created by simulating 5 outages of 2s duration in each of the previously simulated series. For giving an example, the simulated VT measurements of a test run with outages are depicted in figure 3.5.

EKF Model without Biases

In a first experiment, the simulated dataset has been processed with the 9-state EKF model from section 3.3.2. A typical result for a sequence of simulated data with outages is shown in figure 3.6. The result in the figure is displayed as estimation errors for all estimated states, which have been calculated by taking the difference between filter result and ground truth. After an initialization phase, small deviations in position and attitude can be perceived in the estimation result of the filter. However, due to the drift caused by biased IMU

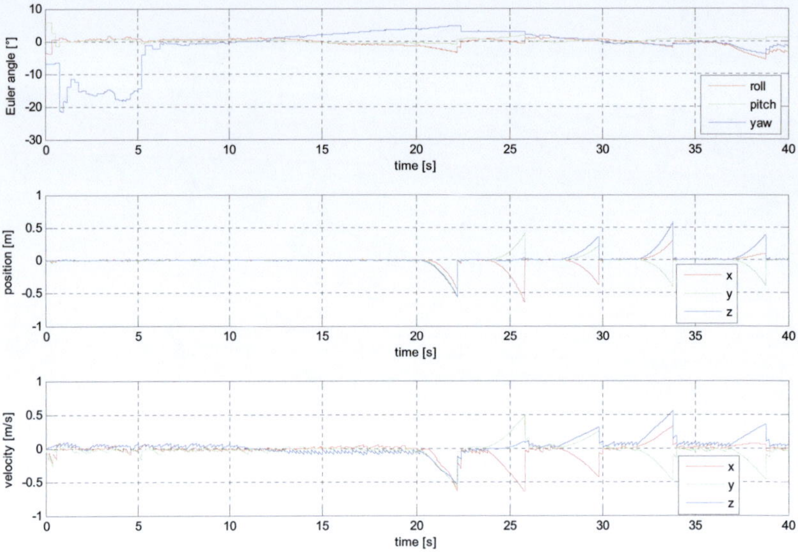

Figure 3.6: Estimation errors of the 9-state EKF model after application to a sequence of simulated data with outages.

measurements, velocity errors and resulting position errors become large during each outage period (position errors in the figure are up to more than half a meter).

The overall performance of the 9-state filter model in all sequences of simulated data is given in table 3.1 in the "EKF without bias" column. The statistical evaluation in the table shows the average and maximal deviations of the filter result to the ground truth over all 10 normal test runs. Additionally, estimation errors calculated over the 10 test runs with simulated outages are shown, as well. Here only the deviation of the last estimation before an outage has been considered for calculating average and maximal deviations. In order to ensure that errors from filter initialization are excluded, the results have been obtained by only considering estimations after 20s in each sequence. The results show that attitude errors are in the range of a few degrees (even after outages). Position errors are typically in the range of several millimeters on average, but can be up to more than 60 centimetres after an outage.

Table 3.1: Performance of the two EKF models in tests with simulated data: Attitude, position and velocity errors. All results are calculated over 10 simulation runs (initialization time of 20s has been considered in each run). The highest values in each row are marked with bold letters.

Filter model		**EKF without bias**			**EKF with bias**		
Euler angle error [°]		ϕ	θ	ψ	ϕ	θ	ψ
normal	average	1.0462	0.5267	**1.9290**	0.0895	0.0569	**0.4509**
	max	3.8752	1.2704	**4.3171**	0.4407	0.5174	**2.5397**
outages 2s	average	2.4352	0.8056	**2.9096**	0.1064	0.0718	**0.6638**
	max	**6.0773**	1.7151	5.8252	0.4200	0.4756	**2.6313**
Position error [m]		x	y	z	x	y	z
normal	average	0.0033	0.0031	**0.0048**	**0.0017**	0.0015	**0.0017**
	max	0.0221	0.0210	**0.0235**	0.0086	0.0074	**0.0165**
outages 2s	average	0.3418	**0.3912**	0.3724	0.0240	0.0183	**0.0250**
	max	**0.6567**	0.6206	0.6151	0.0956	0.0546	**0.1601**
Velocity error [m/s]		v_x	v_y	v_z	v_x	v_y	v_z
normal	average	0.0231	0.0225	**0.0407**	**0.0047**	0.0038	0.0045
	max	**0.1227**	0.1028	0.1097	0.0429	0.0236	**0.0464**
outages 2s	average	0.3960	**0.4497**	0.3619	**0.0224**	0.0166	0.0199
	max	**0.6676**	0.6442	0.5671	0.0946	0.0588	**0.1183**

EKF Model with Biases

In the second experiment, the 15-state EKF model from section 3.3.2 has been applied to the simulated dataset. Estimation errors of the filter result are shown in figure 3.7 for the same series, which has been processed with the 9-state EKF model in figure 3.6. Compared to the 9-state EKF model, the 15-state EKF model is able to yield a smoother result, which has significantly smaller velocity and position drifts in regions of VT outages. The better performance results from the incorporation of bias estimations in the filter models. Angular rate sensor (or gyro) and accelerometer sensor bias estimation errors are shown in figure figure 3.8. In tests with simulated data, bias estimations correctly converge to the simulated offset values (which are

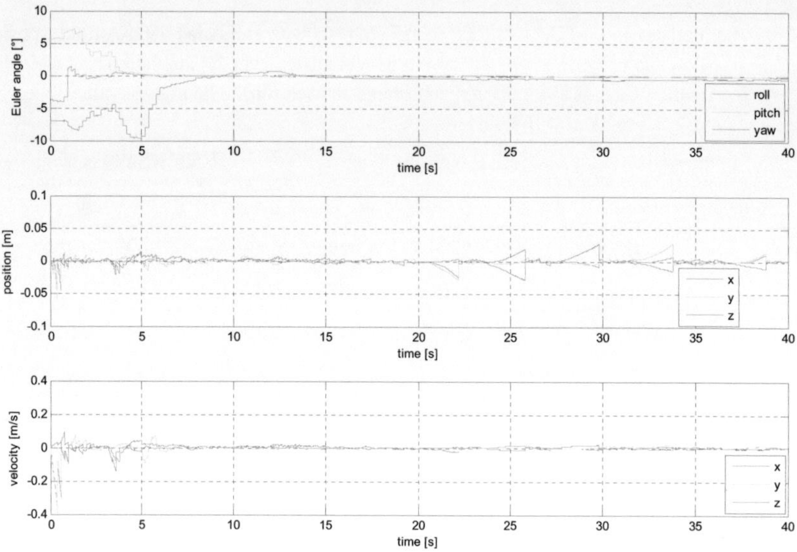

Figure 3.7: Estimation errors of the 15-state EKF model after application to a sequence of simulated data with outages: Attitude, position and velocity.

0.72, 0.15, -0.51$^\circ/s$ for gyros and 0.055, 0.21, -0.17m/s^2 for accelerometers in the simulation).

A statistical overview about the performance of the 15-state EKF model on the simulated dataset is given in the "EKF with bias" column in table 3.1. The evaluation has been conducted after the same manner as with the 9-state EKF model. A comparison of the results shows that the 15-state EKF model is able to yield improved estimation results. Especially, the position error after outages has been reduced to a maximal error of 16 centimeters and average errors in the range of about 2 centimetres (in comparison to more than 30 centimetres for the 9-state EKF model).

The statistical account of bias estimation errors of the 15-state EKF model can be found in table 3.2. Maximal bias estimation errors in the simulation are $0.1^\circ/s$ and $0.1m/s^2$.

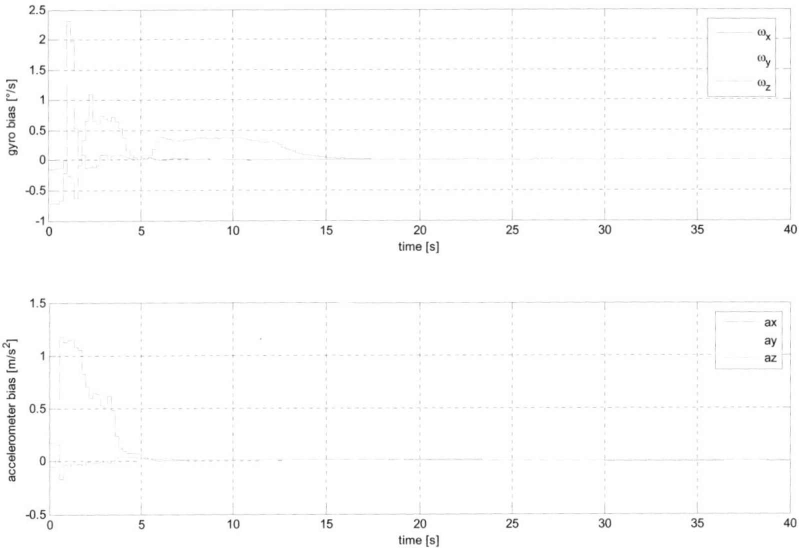

Figure 3.8: Estimation errors of the 15-state EKF model after application to a sequence of simulated data with outages: Gyro and accelerometer biases.

Effect of Synchronization Errors on the EKF Estimation Result

In a third test that has been conducted with simulated data, the influence of synchronization errors on the EKF estimation result has been tested. Synchronization errors can occur, if measurements from the VT system are not correctly synchronized with IMU signals. For the simulation of synchronization effects, a latency of 100 milliseconds has been added to timestamps of the simulated VT measurements. The result after application of the 15-state EKF model to the sequence in figure 3.7 can be seen in figure 3.9. The comparison of the results in figure 3.7 and in figure 3.9 show that the simulated synchronization error has a significant influence on velocity and position estimations. Further tests that have been conducted have shown that a synchronization error in the order of one or two milliseconds (which can be achieved with the hardware synchronization in section 3.2) has no noticeable influence on the EKF estimation result.

Table 3.2: Performance of the 15-state EKF model in tests with simulated data: Gyro and accelerometer bias. All results are calculated over 10 simulation runs (initialization time of 20s has been considered in each run). The highest values in each row are marked with bold letters.

Filter model		EKF with bias		
Gyro bias error [°/s]		$b_{\omega x}$	$b_{\omega y}$	$b_{\omega z}$
normal	average	0.0093	0.0093	**0.0177**
	max	0.0604	0.0192	**0.0936**
outages 2s	average	0.0104	0.0099	**0.0195**
	max	0.0605	0.0206	**0.0902**
Accelerometer bias error [m/s^2]		b_{ax}	b_{ay}	b_{az}
normal	average	**0.0095**	0.0037	0.0082
	max	**0.0914**	0.0155	0.0264
outages 2s	average	**0.0122**	0.0036	0.0085
	max	**0.0903**	0.0151	0.0264

3.4.2. Experiments with Real Data

For testing the position estimation approach with realistic data, experiments have been conducted with a test setup. The workplace dimensions of the test setup are typical for a manual assembly scenario (in the experiments a cube of approximately 1m side length has been considered). In the following, the test setup is described and an account of the VT system performance in this scenario is given. After that, the improvement of position estimations by means of the inertial navigation solution are shown.

Test Setup and Real Dataset

The test setup for recording realistic data is depicted in figure 3.10. In the test setup, the sensor bracelet depicted in figure 3.10 a) is used for tracking the hands of a test person. By means of the sensor bracelet, the IMU is attached to the wrist. For tracking the hand position, VT markers are mounted on the sensor bracelet at known positions in the object coordinate system. The position of the camera in the test setup is illustrated in figure 3.10 b). For testing the position estimation system, a single camera has been placed at

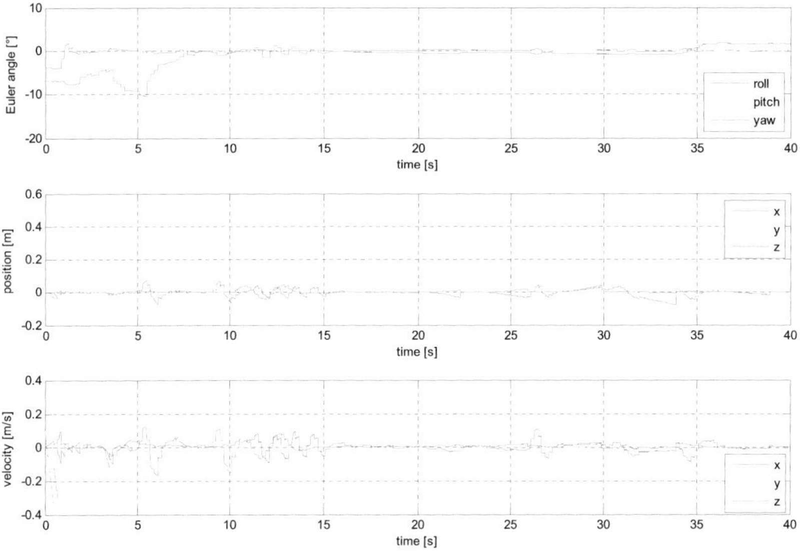

Figure 3.9: Estimation errors of the 15-state EKF model after application to a sequence of simulated data with outages and a synchronization error between IMU and VT data of 100ms: Attitude, position and velocity.

a position from which the working environment could be observed. Before recording, the camera has been calibrated in order to determine the intrinsic and extrinsic camera parameters. Thus, position and orientation of the camera in the world coordinate system are given by the extrinsic camera parameters. The test setup described above has been used for recording 6 sequences of test data, each of 5 minutes duration. All test sequences contain IMU and VT system measurements, which have been recorded while a test person was performing arbitrary hand movements.

VT Performance Tests

When using the VT system for positioning or for updating estimations of the INS solution, it has to be regarded that position measurements are not always available because of outages. Occasional outages result from occlusions, the object being out of the field of view or detection errors of the VT algorithm. Table 3.3 gives an account of the availability of VT measurements in the test data. The results in the table contain information about average sampling

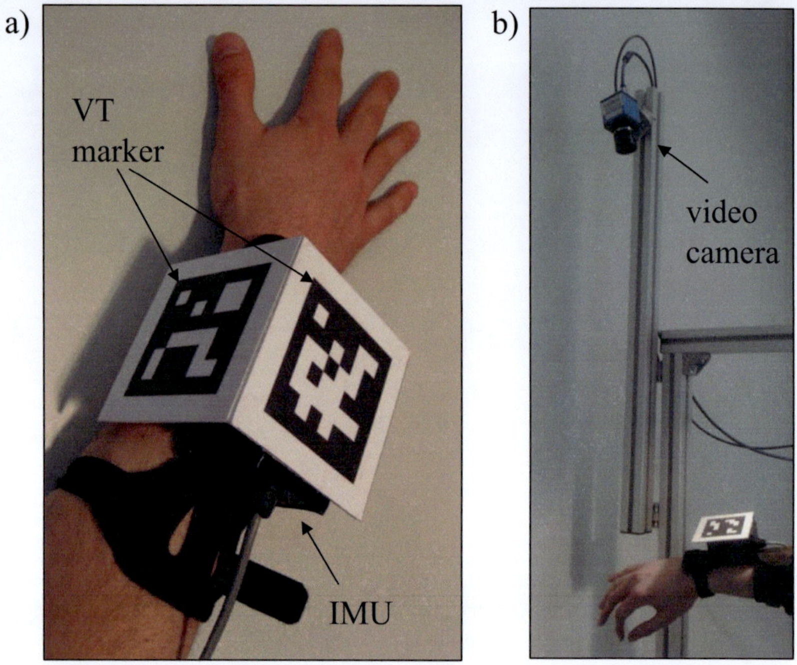

Figure 3.10: Setup for real data tests with the position estimation system: a) Sensor bracelet with IMU and VT markers. b) Camera position for recording hand movements.

rate and quantify outage occurrences in the test sequences. It should be noted that the VT system software achieves maximal frame rates of 8-9 frames per second (fps) on the utilized test PC (Intel core 2 duo, 2.4Ghz, 3.4GB RAM) with a single camera.

Because of the lacking ground truth, the precision of VT measurements could not be evaluated with the test data from arbitrary movements. Therefore, another test has been conducted, in which VT data has been recorded with a single marker that has been moved along paths parallel to the axes of the world coordinate system without performing rotations. The deviations of VT measurements from these known paths, such as deviations of attitude measurements, are shown in table 3.4. In the table it appears that ψ-angle measurements are more precise than measurements of other angles and that z-measurements are more noisy than x- and y-measurements. The reason for

Table 3.3: Availability of VT measurements in the test dataset: Resulting average sampling rate, longest outage and number of outages of long duration.

VT measurements	sampling rate [1]	outage max [2]	#outages >1s [3]
test sequence 1	6.01	2.42	26
test sequence 2	5.86	2.23	18
test sequence 3	6.47	1.62	9
test sequence 4	6.35	2.05	7
test sequence 5	4.42	3.19	49
test sequence 6	5.83	1.94	20
overall	**5.8229**	**2.2405**	**21.5**

[1] average sampling rate in test sequence [fps]

[2] longest outage in test sequence [s]

[3] number of outages >1s in test sequence

this lies in the geometry of the test setup, in which the z-axis of the world coordinate system is parallel to the optical axis of the camera. This causes less noise in ψ-angle measurements and reduces the accuracy of z-measurements. In the utilized setup, the VT system is able to yield position estimations of an accuracy in the range of a few millimeters up to some centimeters in z-direction. The deviation of ψ-angle measurements is usually in the order of a few degrees.

Table 3.4: VT system measurement precision in the test scenario.

Position deviations [m]	x	y	z
average deviation	0.0021	0.0019	0.0191
maximal deviation	0.0122	0.0124	0.1200
Euler angle deviations [°]	ϕ	θ	ψ
average deviation	12.8147	9.0126	1.7489
maximal deviation	62.2740	42.5978	11.2660

EKF Model without Biases

An example for position deviations, resulting from the application of the 9-state EKF model to one of the sequences in the test dataset, is depicted in figure 3.11. Because of the non-availability of a ground truth in tests with

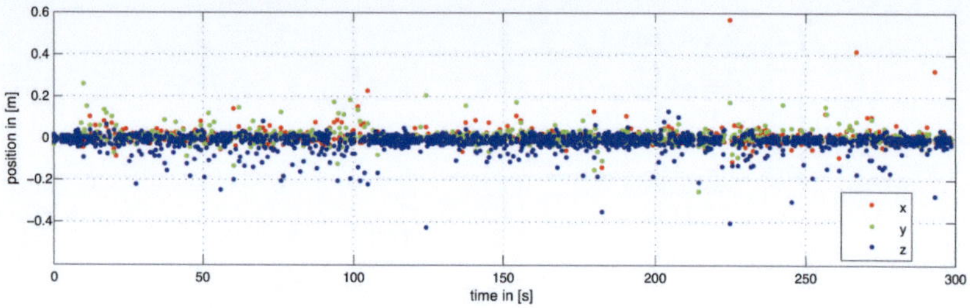

Figure 3.11: EKF result for a real data test sequence: Position deviation of estimations with the 9-state model to VT system measurements at the end of each outage.

real data, VT system measurements have been used as an approximate reference. The position deviations in the figure have been obtained by calculating the difference between the last predicted position before an update and the VT position update. Thus, every point in the figure represents the position drift at the end of an outage.

A statistical overview about all position deviations in the test dataset is given in the "EKF without bias" column in table 3.5. For calculation of the deviations in the table, only position estimation results after 60s have been considered in order to exclude possible errors from filter initialization. The results in the table show that average position deviations are around one centimeter. It should be noted that average z-axis deviations are in the order of average VT z-measurement accuracies. Therefore, no clear statement can be made about average z-axis deviations. Maximal deviations of EKF results to VT results can be up to more than half a meter.

A case that exemplifies the particular usefulness of the INS solution is shown in the cutout of a sequence result in figure 3.12. In the areas, which are marked with dashed circles, fast oscillating movements have been performed. Although the VT system was not able to sample these movements correctly, the original movement has been retrieved to the most extent by means of the INS solution.

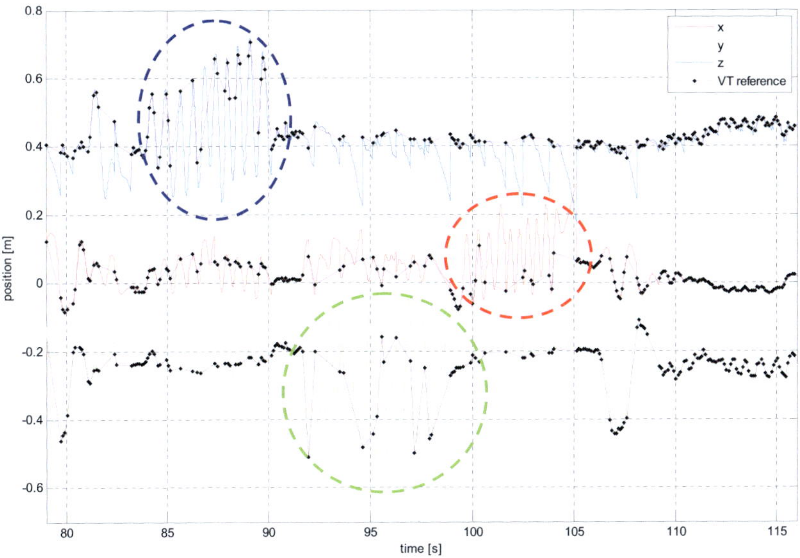

Figure 3.12: Example for lacking VT system position measurements, which are retrieved by the 9-state EKF model. Relevant areas are marked with dashed circles.

EKF Model with Biases

Figure 3.13 shows the deviations of position estimation results from the 15-state EKF model to VT position measurements for the same series, which has been used in figure 3.11. In contrast to the 9-state EKF model, the 15-state EKF model takes about a minute for initialization (which can be seen in the instable deviations at the beginning of the plot). However, the 15-state EKF model seems to be more accurate, because its position deviations appear to be smaller than the deviations of the 9-state EKF model.

The statistical evaluation in the "EKF with bias" column in table 3.5 gives an account of all position deviations of the 15-state EKF model applied to the test dataset. Compared to the results of the 9-state EKF model, the 15-state EKF model is able to yield better results for average and maximal deviation. Average and maximal z-axis deviations of the 15-state EKF model are in the order of VT deviations. As in the case of average deviations of the 9-state EKF model, no clear statement can be made for these results.

Figure 3.14 shows the position result of the 15-state EKF model for the same

Table 3.5: Position deviations of EKF estimations to VT results in the real data test. Results from sequences after an initialization phase of 60s have been considered.

Filter model	**EKF without bias**			**EKF with bias**		
average deviation to VT [m]	x	y	z	x	y	z
test sequence 1	0.0094	0.0122	0.0202	0.0063	0.0075	0.0167
test sequence 2	0.0101	0.0126	0.0199	0.0075	0.0082	0.0137
test sequence 3	0.0090	0.0106	0.0187	0.0069	0.0067	0.0163
test sequence 4	0.0095	0.0096	0.0197	0.0072	0.0064	0.0130
test sequence 5	0.0132	0.0156	0.0269	0.0096	0.0099	0.0221
test sequence 6	0.0110	0.0116	0.0227	0.0078	0.0076	0.0168
overall average deviation	**0.0104**	**0.0120**	**0.0213**	**0.0075**	**0.0077**	**0.0164**
maximal deviation to VT [m]	x	y	z	x	y	z
test sequence 1	0.2021	0.5644	0.5108	0.0747	0.1054	0.1434
test sequence 2	0.3304	0.4557	0.6623	0.1123	0.0846	0.1066
test sequence 3	0.3859	0.2887	0.2454	0.0964	0.1167	0.0890
test sequence 4	0.3278	0.2075	0.8685	0.0884	0.1390	0.0811
test sequence 5	0.4005	0.2411	0.6586	0.1408	0.1778	0.1489
test sequence 6	0.5630	0.2535	0.4258	0.1540	0.1093	0.0998
overall maximal deviation	**0.5630**	**0.5644**	**0.8685**	**0.1540**	**0.1778**	**0.1489**

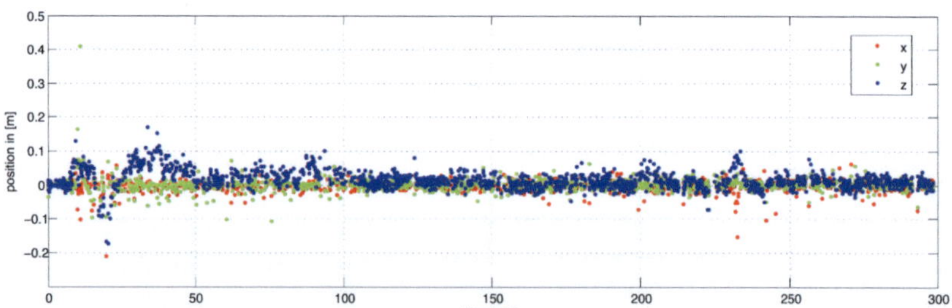

Figure 3.13: EKF result for a real data test sequence: Position deviation of estimations with the 15-state model to VT system measurements at the end of each outage.

sequence part, which has been used in figure 3.12. When comparing the filter results in the figures, it appears that the filter result in figure 3.14 is smoother than in figure 3.12. This indicates that the 15-state EKF model is able to yield better estimations for missing position information in VT outages than the 9-state EKF model.

The bias estimations of the 15-state EKF model for the sequence in figure

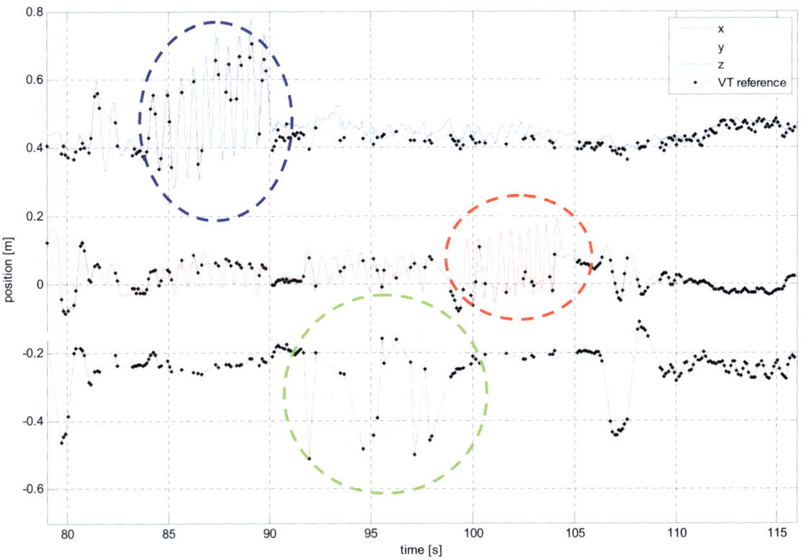

Figure 3.14: Example for lacking VT system position measurements, which are retrieved by the 15-state EKF model. Relevant areas are marked with dashed circles.

3.13 are shown in figure 3.15. After filter initialization, the bias estimations appear to be relatively stable, which indicates a converging behavior.

A final comparison of the maximal filter deviations in table 3.5 and the maximal position errors after outages in table 3.1 reveals that for both filter types, simulation errors are smaller than filter deviations in tests with real data. However, it has to be considered that the VT results (which are used to calculate the deviations in table 3.5) are only an approximate reference and that outages in real data sometimes exceed 2s (see table 3.3). Thus given, the findings from simulation appear to be similar to real data test results.

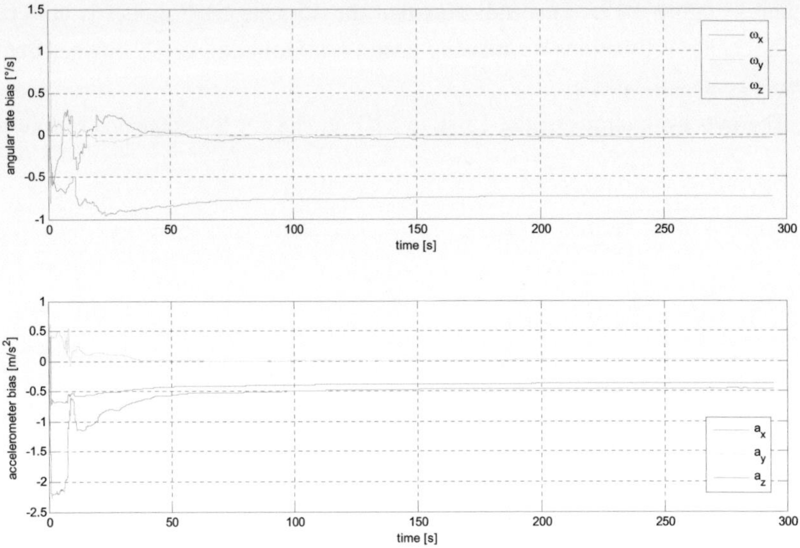

Figure 3.15: Bias estimations of the 15-state EKF model for a sequence of the real dataset. After an initialization time of about a minute, the filter provides a stable result for bias estimations.

3.5. Summary

In this section, a position estimation system has been presented, which is able to determine 3D positions of arbitrary objects in indoor environments. By means of the combination of marker-based VT and inertial navigation, the system has been designed to cover different worker activity recognition scenarios.

The utilization of marker-based VT is appropriate for industrial applications and allows scalability with respect to the dimensions of the tracking area. This means that position tracking can be realized for large scale environments, such as as room-level applications, but also that applications can be covered, which require high tracking accuracy in areas with small dimensions.

Because of the fact, that the position estimation system provides information to activity recognition methods, a gapless tracking result is desirable. Therefore, the drawbacks of outages and limited sampling rates of the VT system have been tackled by utilization of an inertial navigation solution, which is based on a small MEMS IMU and an EKF.

Video-Tracking System

Tests with the marker-based VT system have shown that it is able to yield position estimations with an accuracy up to the order of millimeters in our scenario. Average VT sampling rates basically depend on image frame rates and computing power, but are in practice reduced by outage occurrences. Typical average sampling rates for tracking of hand movements are in the order of 5-6fps.

Inertial Navigation Solution

The presented INS solution utilizes MEMS IMU's and triggerable industrial video cameras, which are operated in a sensor network that provides measurements with synchronized timestamp information. In order to estimate positions from IMU and VT system measurements, two different EKF models have been proposed. The techniques of the INS approach have been tested in a simulation, which allows evaluation of the theoretical performance with respect to an exactly known ground truth. Furthermore, the importance of synchronization and the effect of synchronization errors have been demonstrated in the simulation.

In tests with a realistic scenario it has been shown that the achieved position estimation performance is comparable to simulated results. Short VT outages have been compensated and fast movements have been captured by means of the increased sampling rate that is provided by the INS solution. A comparison of the filter models has shown that the 15-state EKF model takes longer initialization time than the 9-state EKF model. However, the 15-state EKF model is able to yield more accurate results.

4. Low-Level Activity Recognition

In this chapter, our approach for the recognition of low-level activities performed by human workers in industrial environments is presented. As described in section 2.3, the activity recognition approach is partially based on location information of objects involved in the manufacturing process (such as tools or parts of the human body) and cues about performed actions (e.g. fastening a screw). In the component overview of figure 2.2, the techniques of this chapter are represented by the components windowed features generation, action classification, location classification and activity fusion.
After a brief review of sensor devices and processing techniques, which are used in state of the art activity recognition approaches, the utilized feature extraction and action classification methods are explained. Furthermore, an approach for classifying activity related locations and the activity fusion method are presented. Finally, an account of the performance of the presented activity recognition approach is given by an evaluation of experiments with the PC scenario, which has been introduced in section 2.2.

4.1. Introduction

Because of the fact that the recognition of human activities is a wide area of research with different approaches and applications, a variation of sensor devices and processing methods for the recognition of worker activities is existing. In the following, an overview about commonly used sensor devices and activity recognition methods is given and the choice of sensors and processing methods for the worker activity recognition approach of this work is explained.

Sensor Devices

An important category of sensor devices for utilization in activity recognition applications are vision-based systems. Since optical sensors are common for the analysis of human motion [66], vision-based techniques are used in a variety of activity recognition applications, such as the surveillance of office environments [70] or the recognition of sports activities [85]. However, the major drawbacks of vision-based approaches are line of sight problems and the computational complexity of image processing methods.

An alternative to vision-based techniques are wearable sensing techniques. A very popular type of wearable sensors are accelerometer devices, which are utilized for instance in [7] for the recognition of everyday activities. In [99], an assembly task scenario is presented, in which the recognition of worker activities is based on an IMU, force sensitive resistors and an RFID sensor that are attached to the arm of the working person. Over the years, wearable techniques have drawn a lot of interest within the activity recognition community, which led to the development of several sensor platforms. In [62], a miniaturized wearable sensor platform is utilized, which contains accelerometers, light sensors, and microphones. Another wearable sensor platform is presented in [110]. This platform contains accelerometers and tilt switches that are utilized for motion based activity recognition. Physiological information in form of galvanic skin response, skin temperature, near-body ambient temperature and accelerations can be measured with the wearable sensor platform in [55].

Besides the activity recognition approaches that are based on monitoring persons or attaching sensors to persons, an alternative strategy is to attach sensor devices to objects, which are used in activity recognition scenarios. This is for instance demonstrated in the manual assembly scenario of [5], where inertial sensors are attached to workpieces. Another example is the work of [102], which utilizes a combination of worn accelerometers and RFIDs attached to used objects for ADL recognition. A network of distributed binary sensors for the detection of household activities has been utilized in [51]. Although being an approach for simple acquisition of object use information, it has to be mentioned that the attachment of sensors to workpieces or to a large

number locations might be inconvenient for a flexible activity recognition approach.
The combination of position measurements and information from wearable devices is an approach to increase the performance of activity recognition applications, in which locations are related to activities. An example for this is given by the work of [103], where information from different wearable devices is utilized in combination with position information from GPS measurements for activity recognition in outdoor environments. In [97] an approach for indoor recognition of maintenance activities is presented, which is based on an ultrasonic position tracking system and inertial sensors.

Activity Recognition Methods

A widely used technique for the recognition of activities in sequences of sensor data is to calculate particular features (neglecting the phase relations) from segments of windowed raw data and to apply a pattern recognition method [44] in order to classify performed activities from calculated features. Thus, generalized information is derived from the spatio-temporal content (e.g. the course of a trajectory) of the signal segment in each window. In [7] this approach has been demonstrated with accelerometer signals. After feature calculation from raw data in a sliding window, a decision table method, a nearest neighbor classifier, a decision tree and a naive Bayes classifier have been applied and the results have been compared. Due to the popularity of this approach, a number of similar applications can be found in literature. For instance the work of [81], in which the performances of several base-level classifiers and meta-level classifiers are compared.
For activity recognition with vision-based sensors, often generative models are utilized [15]. One of the most commonly used generative modelling method for vision-based activity recognition are hidden Markov models (HMM) [70, 72]. However, HMMs find also application in activity recognition with wearable sensors [48, 126]. The application of HMMs to activity recognition is useful if activities can be modeled as time series patterns, which are for instance present in gestures. Therefore, this approach is to some extent in contrast to the generalizing sliding window method. Another approach for the classification of activities with underlying trajectory information is the usage of template matching methods, e.g. as demnonstrated in [98].

There are also activity recognition approaches existing, in which the results from sliding window techniques and generative models are combined. An example for this is given in [115], where a hidden Markov model is applied to accelerometer readings and sound information is classified with a linear discriminant analysis from microphone data.

As shown in the component overview in figure 2.2, the activity recognition approach presented in this work is based on the recognition of worker actions and on object location information. Location information constitutes a useful cue with respect to performed activities, which has also been demonstrated in other work (e.g. in [103]). This is especially the case for the structured working environments of modern factories. For acquiring position information about objects that are involved in the manufacturing process, the system for indoor position estimation in industrial environments that has been presented in chapter 3 is utilized.

The recognition of actions is based on motion data from IMU measurements (which is an already existing component of the position estimation system). The inertial sensors of a small MEMS IMU have various advantages, which makes them suitable for our approach. Inertial sensors have proven to provide useful physical information for activity recognition in several applications. Furthermore, they are of a relatively small size and can be attached to tools or clothing. Finally, inertial sensors are robust to external disturbances and are therefore adequate for industrial environments.

For the recognition of activities a combined approach is used, which is based on action classification with a sliding window technique and density based location classification. This approach is explained in detail in the following sections.

4.2. Action Classification with Wearable Sensors

Since it can not be expected that worker actions are generally performed in a way that they show well defined temporal patterns, an approach based on the generalizing sliding window method proposed in [7] is utilized for action

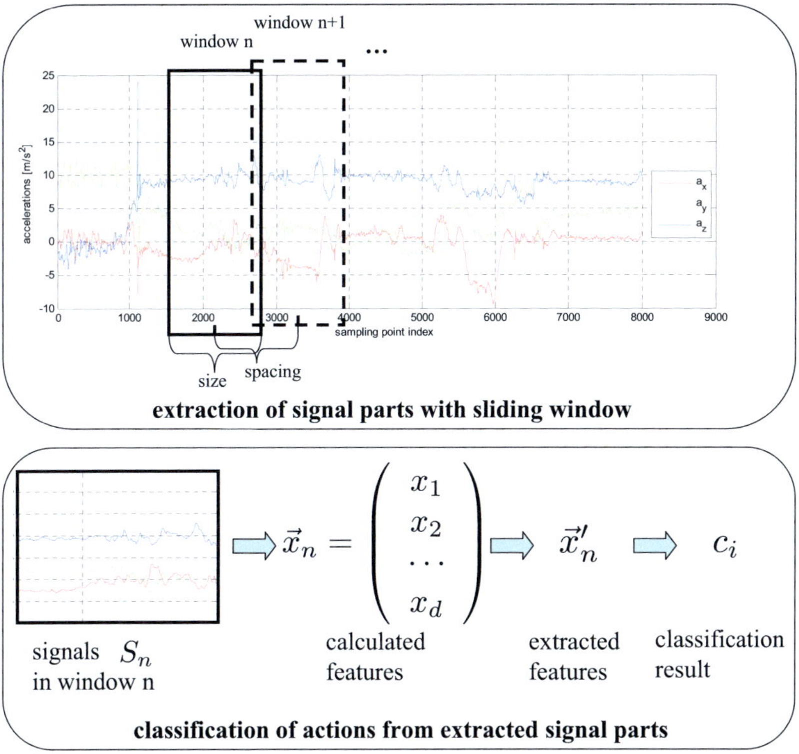

Figure 4.1: Classification of actions from sensor signals based on a sliding window. Top: Segmentation of recorded sensor data by means of a sliding window. Bottom: Classification of extracted segments.

classification in this work. The generalizing effect is able to cope with variations in the execution of actions within the sensor data of a single subject such as between sensor data from multiple subjects.

The approach for action classification based on sliding a window is depicted in figure 4.1. Assuming that we are receiving a sequence of data from one or more sensor devices that are attached to particular positions at the worker or at an object used by the worker, signal parts S_n are extracted for every sensor by means of the sliding window illustrated at the top of figure 4.1:

$$S_n = (s_1, s_2 \dots s_L). \tag{4.1}$$

The length in sampling points L and the displacement of extracted information result from the window size and the window spacing parameters, which are to be chosen according to the duration of performed actions or the duration of characteristic action parts. From extracted signal parts, features (e.g. average value or signal energy) are calculated resulting in a feature vector $\vec{x}_n$, as shown at the bottom of figure 4.1. Since calculated features are often of a high dimension, it is recommendable to apply a feature extraction method in order to obtain a feature vector of reduced dimensionality $\vec{x}'_n$ containing most of the relevant information of $\vec{x}_n$. The extracted feature vector $\vec{x}'_n$ can then be used as input information for the classification of action classes c_i with an adequate classification method.
Basically, a variety of statistical learning methods are existing [21, 57, 86, 119], which are suitable for solving different types of classification problems. However, since the performance of a classifier always depends on the type of the classification problem (e.g. distribution of the data, size of training dataset, etc.) and since our intention is to stay flexible with respect to the application, two standard methods have been chosen, which find application in many activity recognition applications. The first choice was for the naive Bayes method, because it is a very common classification technique and is often used as benchmark for other classifiers. As an alternative method, the k-nearest neighbor (k-NN) algorithm has been chosen, since it usually yields good results in activity recognition problems. Furthermore, with the naive Bayes and the k-NN method a parametric and a non-parametric classification technique have been selected.

4.2.1. Feature Calculation

The selection of adequate features is crucial to the performance of a classification method in every classification problem. In [42] a study on the selection of features from accelerometer signals for activity recognition has been undertaken, which concluded with the fact that the appropriateness of features generally varies with the utilized scenario. This means that, theoretically, with every change of application parameters, such as the actions to be recognized or the placement of sensor devices, a different combination of features can lead to good classification performances. However, since a careful selection of appropriate features in every scenario stands in contrast with a flexible approach, exhaustive feature studies are not feasible for the purpose of this work.

The approach in this work is based on the selection of a number of initial features that are commonly used in state of the art activity recognition systems reviewed in section 4.1. After the calculation of these initial features, relevant features are derived in a post processing step by means of a feature extraction method. For the recognition of actions with angular rate sensors and accelerometers the following initial features have been used:

- mean: Calculating the sample mean $mean(S_n) = \frac{1}{L}\sum_{l=1}^{L} s_l$ of a windowed signal is one of the most commonly used features, since it is utilized for action classification in almost every wearable sensor application (see [7, 42, 62, 81, 115]). This feature contains the static component of the signal part in the extraction window.
- median: Although the median constitutes a measure that is similar to the mean value, it is used as an additional feature in this work because of it robustness to outliers.
- local extrema: Local extrema are determined by taking the minimum $min(S_n)$ and the maximum $max(S_n)$ of the extracted signal part (e.g. used in [110]). Thus, these features contain information about the maximal intensity of a signal.
- variance: The sample variance $\frac{1}{L-1}\sum_{l=1}^{L}(s_l - mean(S_n))^2$ is a feature for measuring the dynamic content of a signal part. In some publications (e.g. in [7, 42, 62, 81]) the signal energy is used for measuring the dynamics of actions. However, it has to be noted that the variance is proportional to the signal energy if the DC component (i.e. the mean) is removed in the calculation of energy.
- frequency domain entropy: The frequency domain entropy (used in [7] or [42]) is a measure for the average information content of the signal in the frequency domain. This means that it can be used to discriminate actions with similar signal energy, but different energy spectra. The frequency domain entropy can be calculated from the signal in the

frequency domain s_f according to $-\sum_{f=1}^{N} s_f \cdot ld\left(s_f\right)$.

- correlation: A method to determine the similarity of different signals is the correlation measure. The correlation of the two signals $S_{n,1}$ and $S_{n,2}$ is calculated by $\frac{E((S_{n,1}-E(S_{n,1}))(S_{n,2}-E(S_{n,2})))}{std(S_{n,1})std(S_{n,2})}$, where $E()$ is the expected value and $std()$ is the standard deviation. Applied to 3D accelerometers (or angular rate sensors), the similarity of 3 pairs of axis signals (x,y, x,z and y,z) can be determined [7, 42, 81].

All initial features are calculated for each sensor signal of the utilized IMU. Considering that the IMU provides measurements from 3 orthogonal accelerometers and 3 orthogonal angular rate sensors the total sum of calculated features is 42.

4.2.2. Feature Extraction

A problem of high feature dimensions is that they can cause overfitting problems in classifier training if the available training data is limited. On the other hand, it is usually the case that not every feature contains the same amount of information in a classification problem. Thus, one approach to reduce the feature dimensionality is to combine features that contain similar information. This approach is also referred to as feature extraction.

In this work a principal component analysis (PCA) has been used for reducing the dimensionality by linear combination of redundant features. The PCA is a widely used method for feature extraction and reduces the feature dimension by finding the principal components in the input data, that is the axes in the d-dimensional feature space, which optimally represent a set of data in a least-squares sense. The dimension of the input data can then be reduced by projection on the d' principal components (with $d' < d$), which represent most of the information. This principle is exemplified for two dimensions in figure 4.2. It can be seen in the example of the figure that most of the information of the distribution is represented by the principal component indicated by the green arrow.

Given a set of input data in the form of N samples of feature vectors $\vec{x}_n$ of dimension d, the principal components can be found by determining the eigen-

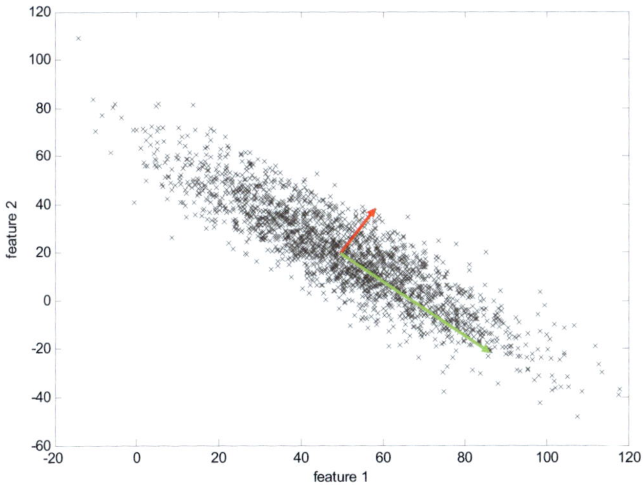

Figure 4.2: Principal components (displayed as arrows) of a set of data: an example with 2 feature dimensions.

vectors $\vec{e}_k$ and the eigenvalues of the scatter matrix ([21]):

$$S = \sum_{n=1}^{N} (\vec{x}_n - \vec{m})(\vec{x}_n - \vec{m})^T, \tag{4.2}$$

where $\vec{m}$ is the sample mean of the dataset. The new feature vectors $\vec{x}'_n$ of reduced dimension d' can then be calculated by the projection

$$\vec{x}'_n = \vec{m} + \sum_{k=1}^{d'} a_{kn}\vec{e}_k. \tag{4.3}$$

The coefficients a_{kn} in equation (4.3) are given by:

$$a_{kn} = \vec{e}_k^T (\vec{x}_n - \vec{m}). \tag{4.4}$$

For the reduction of the feature dimension only the eigenvectors with the d' highest eigenvalues are utilized in equation (4.3).

4.2.3. Naive Bayes Classifier

The naive Bayes method is a statistical classification approach, which is based on the assumption that a mathematical model of the decision problem in form of probabilities and probability densities is existing. In the following, the naive Bayes approach should briefly be explained (for further information on Bayesian decision theory and naive Bayes classification see also [21] or [57]).

A typical target in classification problems is to obtain the posterior $P(c_i|\vec{x})$, which represents the probability that a measured feature vector $\vec{x}$ belongs to a certain class. The posterior can be used for classification after the maximum a posteriori (MAP) method, which is to decide for the class with the highest posterior. However, since the posterior can usually not be determined directly, it is calculated in Bayesian decision theory by making use of the Bayes formula:

$$P(c_i|\vec{x}) = \frac{p(\vec{x}|c_i)\,P(c_i)}{p(\vec{x})}. \tag{4.5}$$

Besides the evidence $p(\vec{x})$, the Bayes formula in equation (4.5) contains the likelihood $p(\vec{x}|c_i)$ and the prior $P(c_i)$. According to the naive Bayes approach, the likelihood can be determined from the feature probability densities $p(x_k|c_i)$ according to

$$p(\vec{x}|c_i) = \prod_{k=1}^{d} p(x_k|c_i), \tag{4.6}$$

if the features are conditionally independent. The prior in equation (4.5) can be derived from knowledge about the occurrence of classes, for instance from a set of training data.

Disregarding the evidence $p(\vec{x})$, which is independent of the class, the MAP decision rule, the Bayes formula and equation (4.6) lead to the following decision rule:

$$c_{NB} = \arg\max_i \left(\prod_{k=1}^{d} p(x_k|c_i)\,P(c_i) \right). \tag{4.7}$$

In practical applications it is often assumed that the feature probability densities are normally distributed. In this case, they can be estimated by calculation of the sample mean and the sample variance of the feature from a set of training data. This feature model is also used for naive Bayes classification in this work.

It should be mentioned that the naive Bayes method is known to yield good classification results even if the features are not conditionally independent [21].

4.2.4. k-Nearest Neighbor Classifier

In contrast to the naive Bayes method, the k-NN approach makes no basic assumptions about the nature of the distribution of features. The idea behind the k-NN method is a quite simple one. Assuming that a set of training data is given that contains enough training samples (or instances) to describe the class distributions in the feature space, the class membership of a test instance is determined from the dominating class of the training samples in the neighborhood of the test instance. The neighborhood is represented by the k closest training instances according to a chosen distance measure. As distance measure between two point vectors $\vec{p}_1$ and $\vec{p}_2$ often a metric of the class of L_k norms is used:

$$L_k(\vec{p}_1, \vec{p}_2) = \left(\sum_{i=1}^{d} |\vec{p}_1 - \vec{p}_2|^k \right)^{1/k} . \tag{4.8}$$

In this work a k-NN classifier is used, which is based on the L_2 or Euclidean norm.

An example for the k-NN classification principle is given in figure 4.3. For k=1, the closest instance is of class 2. Therefore, we would decide in this case that the test point belongs to class 2. However, the instance close to the test point could also be an outlier, since the density of class 2 is quite sparse in this area. This potential misclassification can be avoided by choosing k>1. If k is chosen to be 3, the 3 closest instances to the test point are determined and the class affiliation of the test point is assigned according to the class with the highest occurrence (which would in this case be class 1). Potential class ambiguities can be resolved by additional incorporation of the average distance.

As previously mentioned, one of the advantages of the k-NN method is that no specifications of the class distributions in the feature space have to be made. This can be helpful if the class distributions vary with performed actions or the number of test persons. Despite of its simplicity, the k-NN

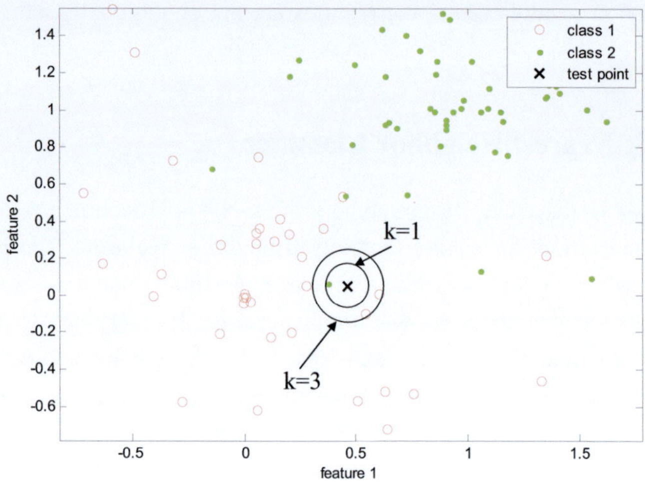

Figure 4.3: Nearest neighbor classification example with 2-dimensional features. The circles around the test point mark the distances that are considered for classification if $k = 1$ and $k = 3$ when utilizing the Euclidean norm.

method is able to yield a high classification quality [21]. A drawback of the k-NN method is that its computational costs increase with the number of instances in the training data, although there are efficient search strategies existing (see also [21] and [119]).

4.3. Position Based Location Classification

The intention for the incorporation of location information in the worker activity recognition process is that the presence of particular objects (e.g. a human hand or a tool) at defined locations is linked to the execution of activities. For obtaining location information, a classification approach is required, which relates position measurements to locations that are relevant for performed activities.

Our general approach to this problem is to treat locations as position densities, which can be derived from a set of training data. As an example, a set of hand position measurements is given in figure 4.4. The measurements have been taken with the VT system from section 3.2.3 while a test person was

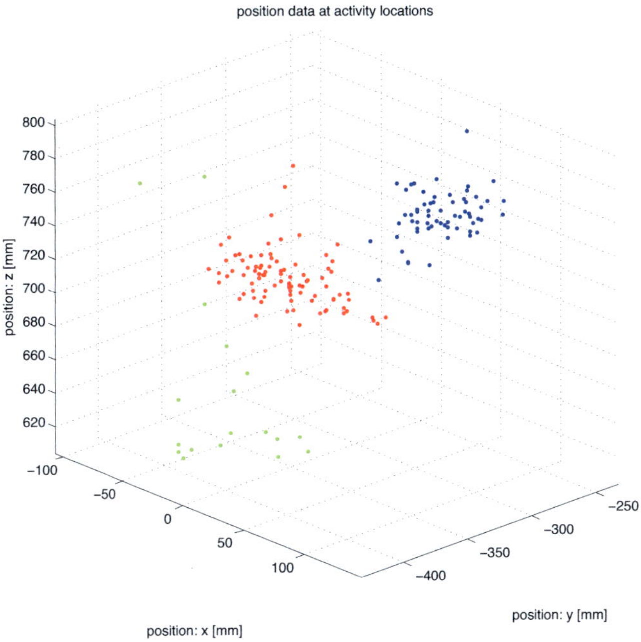

Figure 4.4: Hand position measurements recorded during the execution of three different activities of a manual assembly task. The position measurements appear as point clouds belonging to different location classes, which are indicated by the colors red, green and blue.

performing three different activities of a manual assembly task. The activity locations are represented by point clouds of position samples.

Similar to the action classification techniques in section 4.2, a parametric and a nonparametric approach for location classification based on position densities are proposed. The parametric approach is based on maximum likelihood classification by modelling point clouds as multivariate normal distributions. The nonparametric alternative for being independent from assumptions about underlying distributions is the instance-based k-NN classification method.

4.3.1. Maximum Likelihood Approach

A natural approach to the location classification problem is presented in [97], where a threshold method based on the Mahalanobis distance is used to classify locations. The approach utilized in our work is more general in the way that location classes c_j are related to positions $\vec{p} = (x, y, z)^T$ by means of probability distributions $p(\vec{p}|c_j)$. Thus, the location class of a position measurement can be classified after the maximum likelihood (ML) method according to:

$$c_{ML} = \arg\max_j \left(p\left(\vec{p}|c_j\right)\right). \tag{4.9}$$

In this work, the normal distribution is used to model the probability distributions of positions $\vec{p}$ in 3D space:

$$p\left(\vec{p}|c_j\right) = \frac{1}{(2\pi)^{3/2}\sqrt{|\Sigma_j|}} e^{-\frac{1}{2}(\vec{p}-\vec{\mu}_j)^T \Sigma_j^{-1} (\vec{p}-\vec{\mu}_j)}. \tag{4.10}$$

After the determination of the distribution parameters μ_j and Σ_j from a set of training data by means of the sample mean and the sample covariance, the following formula can be used for location classification:

$$c_{ML} = \arg\max_j \left(\frac{1}{(2\pi)^{3/2}\sqrt{|\Sigma_j|}} e^{-\frac{1}{2}(\vec{p}-\vec{\mu}_j)^T \Sigma_j^{-1} (\vec{p}-\vec{\mu}_j)}\right). \tag{4.11}$$

However, when using the ML approach for location classification it has to be considered that also position measurements, which are far away from the position distribution centers are assigned to location classes. In order to avoid these unrealistic assignments, a position measurement is assigned to a null class, if the probability for each class is below a threshold value. This threshold value can automatically be determined by calculating the density value at the multiple of a standard deviation σ around the normal distribution center. An illustration of possible thresholds in form of 2-dimensional projections of hyperellipses around the distribution centers onto the $x-y$, $x-z$ and $y-z$ planes is given in figure 4.5 for the position measurements of figure 4.4. The ellipses in the figure represent the 1σ, 2σ, 3σ isolines.

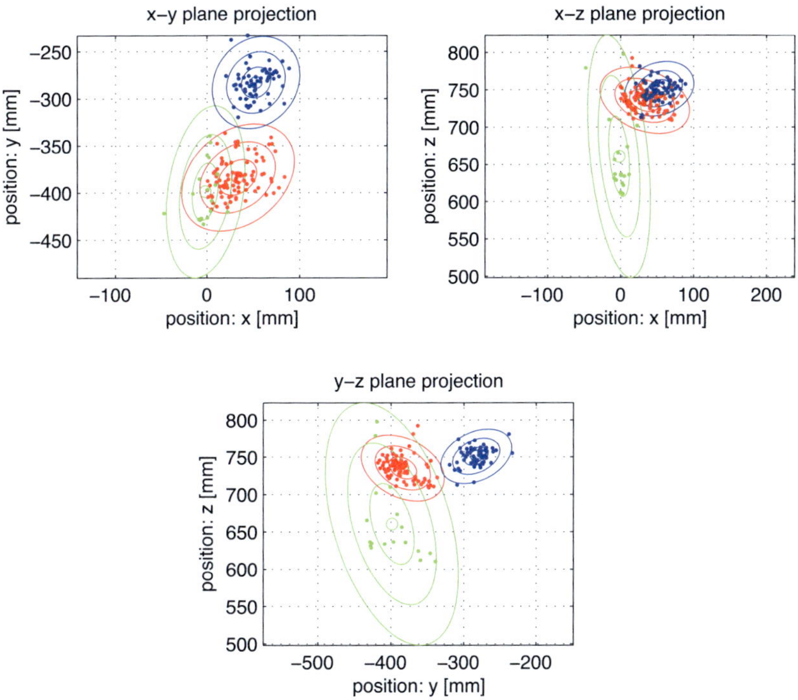

Figure 4.5: 2-dimensional projections of the position measurements of figure 4.4 onto the $x-y$, $x-z$ and $y-z$ planes. The 1σ, 2σ, 3σ ellipses enclose possible confidence areas.

4.3.2. k-Nearest Neighbor Approach

Although the assumption that locations are related to normally distributed position measurements represents usually a good approximation, this must not be the case for every activity. An example for this would be an activity, in which movements occur in the form of trajectories that are not conform with the normal distribution. Furthermore, it happens occasionally that activities are executed by different workers in different manners, e.g. because they hold a tool in another way.

The idea to tackle this problem is to make use of the instance-based learning principle of the k-NN method that has been described earlier in section 4.2.4.

The location class of a position $\vec{p}$ is then classified after the mostly occurring class among the k closest position instances given by a set of training data.

4.4. Activity Fusion

Since the recognition of activities is based on action and location information, the results from the action and location classification methods have to be fused in order to derive activities. Therefore, detected actions and location information first have to be temporally aligned and then fused by an adequate classifier fusion technique for online activity recognition.
For temporal alignment, pairs of location classification and action classification results are created by taking each classified location and determining the temporally closest action classification result. Depending on the utilized task recognition technique, the aligned classification results are either fused in order to derive activities or passed directly to the task recognition level (for task level activity recognition see chapter 5).
For the fusion of classification results different methods are existing. However, the applicability of classifier fusion techniques mainly depends on the type of the output information that is produced by the classification methods. According to [87], the type of a classifier output can be divided into crisp labels (i.e. only the classified class), class rankings and soft outputs (e.g. a score value for each class). Depending on utilized action and location classification techniques, either soft outputs (naive Bayes and ML method) or crisp labels (k-NN method) are available for fusion. In order to cover all types of output information that may be produced by classification, a majority voting technique has been utilized, which is similar to the plurality voting method used in [81]. This means for the fusion of action and location classification results that an activity is considered as detected if both results correspond to the respective activity and otherwise a null class is assigned.

4.5. Experiments and Results

In this section an account of the performance of the presented worker activity recognition approach is given by means of evaluations that are based on experiments with a PC assembly scenario. After a description of the scenario and the experiments, evaluation results for action and location classification such as activity recognition are shown and compared in the following.

4.5.1. PC Scenario

The PC scenario, which has briefly been introduced in section 2.2, represents typical manual assembly tasks of the electrical industry. It consists of a test setup in the form of a laboratory workplace environment, in which a worker assembles a PC from several components.

Workplace Environment and Worker Task Description

An overview about the laboratory workplace environment is given in the image on the left of figure 4.6. This test environment represents a structured industrial workplace, in which workpieces and components are located at defined storage places. The storage places of the mainboard, the CD drive and the screws are indicated in the image at the top right of figure 4.6. The PC case in the middle of the image is placed at a defined position as well, which is not changed during the execution of a task.
The worker task consists of 4 different (sub-)tasks that represent the main steps in the PC assembly. These tasks are opening the case by removing the cover plates, assembling the CD drive and assembling the mainboard in the empty case shown at the bottom right of figure 4.6, such as connecting the power and data connectors.
In the development phase of the activity recognition system, a number of activities that are essential for the recognition of each worker task have been identified. These activities are listed together with their corresponding locations and actions in table 4.1. The result of the assembled components after the completion of the whole worker task and some of the locations, at which activities are performed are illustrated in figure 4.7.

Table 4.1: PC scenario: List of tasks and activities with corresponding locations and actions. Activities are listed in the order of their execution (CD = CD drive, MB = mainboard).

task		**activity**		**location**		**action**	
#	description	#	description	#	description	#	description
1	remove cover plates from case	1	remove screw 1	1	case cover screw 1	1	use screwdriver
		2	remove cover 1	2	case cover 1	2	pull up
		3	remove screw 2	3	case cover screw 2	1	use screwdriver
		4	remove cover 2	4	case cover 2	2	pull up
2	assemble CD drive	5	pick up CD	5	CD storage place	3	grasp
		6	pick up CD screw	6	CD screw box	3	grasp
		7	fasten CD screw 1	7	CD screw 1	1	use screwdriver
		6	pick up CD screw	6	CD screw box	3	grasp
		8	fasten CD screw 2	8	CD screw 2	1	use screwdriver
		6	pick up CD screw	6	CD screw box	3	grasp
		9	fasten CD screw 3	9	CD screw 3	1	use screwdriver
		6	pick up CD screw	6	CD screw box	3	grasp
		10	fasten CD screw 4	10	CD screw 4	1	use screwdriver
3	assemble mainboard	11	pick up MB	11	MB storage place	3	grasp
		12	pick up MB screw	12	MB screw box	3	grasp
		13	fasten MB screw 1	13	MB screw 1	1	use screwdriver
		12	pick up MB screw	12	MB screw box	3	grasp
		14	fasten MB screw 2	14	MB screw 2	1	use screwdriver
		12	pick up MB screw	12	MB screw box	3	grasp
		15	fasten MB screw 3	15	MB screw 3	1	use screwdriver
		12	pick up MB screw	12	MB screw box	3	grasp
		16	fasten MB screw 4	16	MB screw 4	1	use screwdriver
		12	pick up MB screw	12	MB screw box	3	grasp
		17	fasten MB screw 5	17	MB screw 5	1	use screwdriver
4	plug in connectors	18	connect CD data	18	CD data connector	4	plug in
		19	connect CD power	19	CD power connector	4	plug in
		20	connect MB power	20	MB power connector	4	plug in

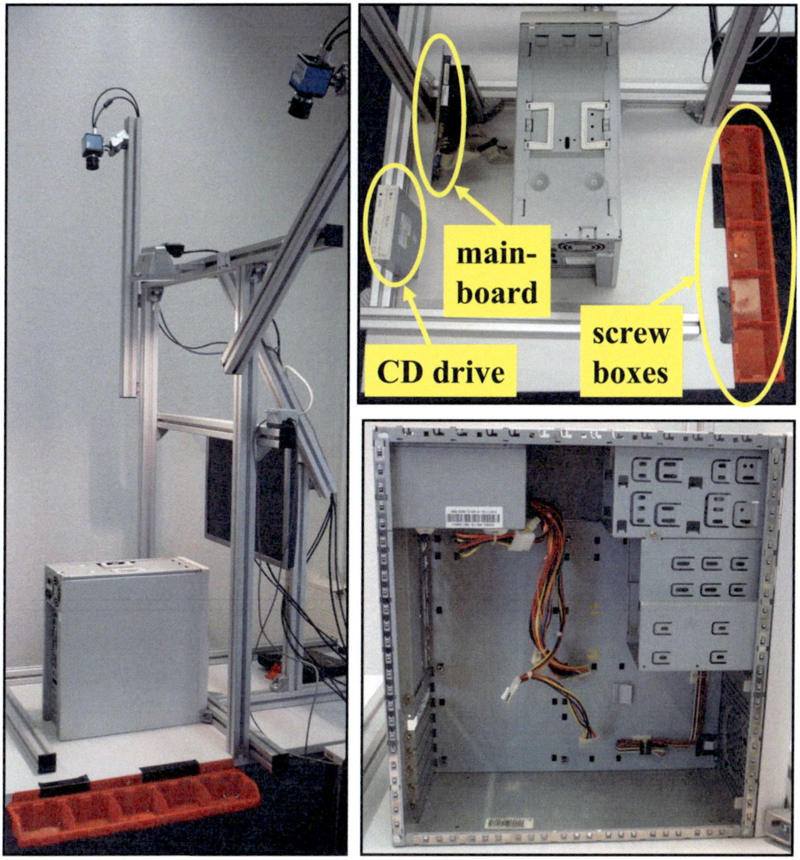

Figure 4.6: PC scenario: Left: Laboratory workplace environment overview. Right (top): PC case position and storage places of the mainboard, the CD drive and the screws. Right (bottom): Empty PC case

Since motion and position of the worker's hands constitute the most important source of information in manual assembly, position and IMU measurements that are taken with the sensor bracelet from section 3.4.2 provide the physical input data to activity recognition in the PC scenario. The VT system setup consists of two video cameras, which have been placed above the workplace (see also the left image of figure 4.6). Although the marker-based position tracking system needs only one camera for measuring 3D positions,

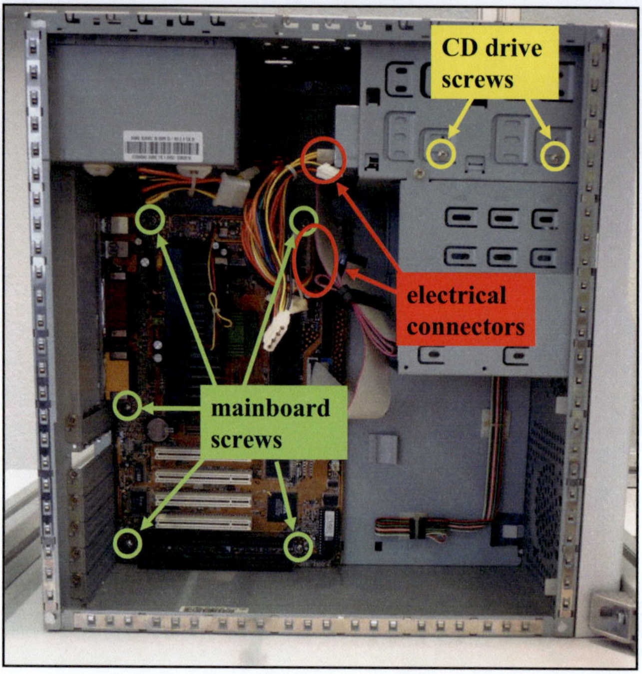

Figure 4.7: PC scenario: PC case with assembled components and locations of mainboard screws (screw 1-5), CD drive screws (screw 1+2 only, screw 3+4 are on the other side) and electrical connections (CD data connector is behind the CD power connector).

two cameras are used in this scenario for increasing the tracking accuracy and for a better observation of the workplace environment.

Test Dataset and Evaluation Procedure

As a basis for evaluation, a test dataset has been recorded with 6 participants. The dataset consists of sequences, in which a participant is performing the assembly tasks that are listed in table 4.1. The 4 tasks have been performed 6 times by all participants. Thus, 36 sequence instances consisting of data from 6 participants have been recorded for each task. During the performance of assembly tasks, position data and IMU measurements have been recorded with the sensor devices and ground truth (GT) labels, which mark the start and the end of each executed activity, have been assigned manually. For sim-

plification of the labelling and evaluation processes, all activities have been performed primarily with the right hand (the left hand has only been used for supporting the right hand). However, it should be noted that the system setup and the proposed techniques are generally not restricted to the recognition of activities that are performed with one hand only.
The described action and location classification methods have been trained with samples that have automatically been cut out from recorded data by using the GT labels as a reference. After training, the methods for the recognition of actions, locations and activities have been applied to the sequence data and the results have been evaluated with the GT information. All tests have been conducted by means of a threefold cross validation (CV) with mixed data from all participants. This means that for each task 24 sequence instances have been used for training and the remaining 12 for testing. In order to test all 36 sequence instances, this procedure has been repeated three times.
An example for the results from action and location classification and activity recognition is given in figure 4.8. Since a sliding window technique is used for action classification, detected actions appear in the figure as areas spanning a length of the window width. Location and activity detections are marked as points and occur at the time of the corresponding position measurement. For the evaluation of the produced test results, true positives (TP) and false positives (FP) are counted under consideration of the GT according to:

- TPs are defined as detections that overlap with the GT area of the respective class. For location and activity detections a tolerance of half a second is added to the start and the end of manually assigned labels and detections, which occur directly before and after a TP of the same class are disregarded. Multiple detections at the same GT label are considered as one TP and are not treated specifically.
- Detections that remain after the identification of TPs are counted as FPs. Multiple action detections overlapping with a FP of the same class are disregarded. Multiple location and activity FPs that occur directly before and after a FP of the same class are disregarded, as well.

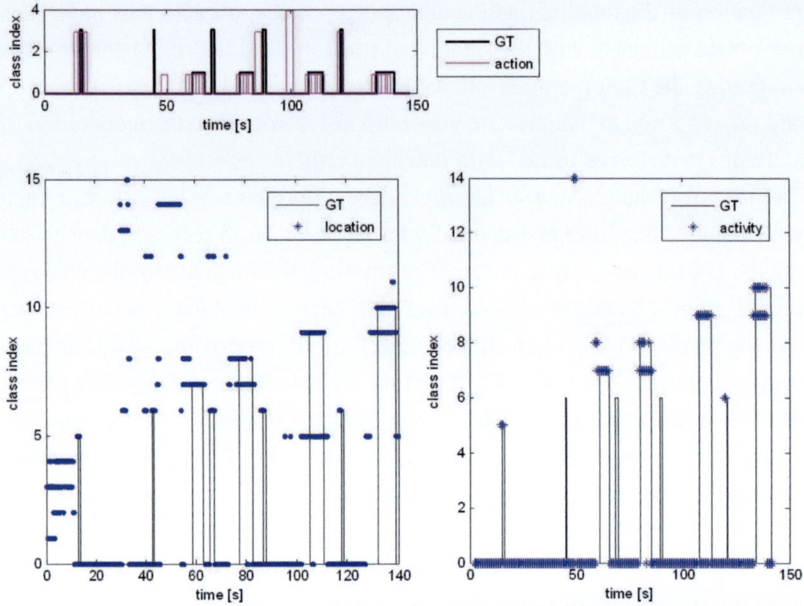

Figure 4.8: Result examples for action classification (top), location classification (bottom left) and activity fusion (bottom right). The test sequence, which has been used in this example, is of task class number 2.

Based on TP and FP detections the precision rate

$$precision = \frac{TP}{TP+FP} \tag{4.12}$$

and the recall rate

$$recall = \frac{TP}{TP+FN} \tag{4.13}$$

are determined as measures for evaluation. It should be noted that in this work good recall rates are more desirable than good precision rates for low-level activity recognition. The reason for this is that low-level activities form the input for task recognition, as shown in figure 2.2. Thus, false alarms can be reduced on the task recognition level to some extent by the incorporation of context knowledge. However, missing detections cannot be retrieved in any further step.

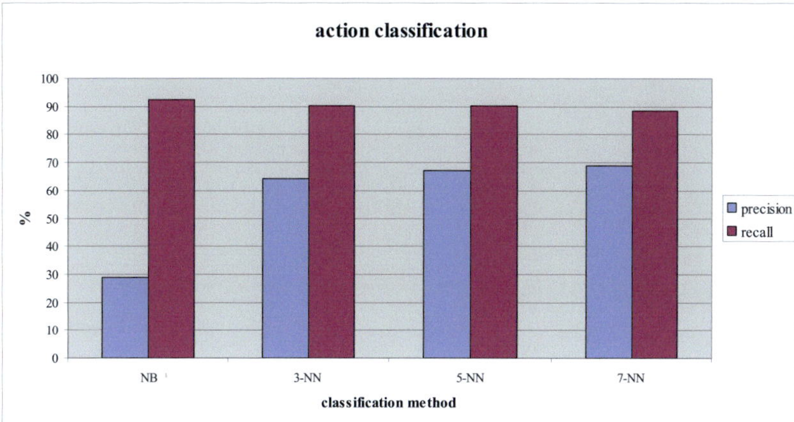

Figure 4.9: Action classification results of the naive Bayes (NB) and the k-NN classifier (for k=3,5 and 7) in the PC scenario (recognition rates are averaged over all classes).

4.5.2. Action Classification with Wearable Sensors

Figure 4.9 shows the results from action classification with the naive Bayes (NB) and the k-NN methods applied to the IMU test sequences of the PC scenario. In all tests, a window size of 1024 sampling points (about 2.5 s) and a window spacing of 512 sampling points (about 1.25 s) have been used, which yield comparatively good results for this data set. Before classification, a PCA has been applied for feature extraction.

The results show that, compared to the NB method, the k-NN classifiers yield slightly reduced recall rates but significantly better precision rates. A possible explanation for this is that the execution of actions often varied from participant to participant, which has been observed during recording. Thus, the instance based k-NN method is able to yield better results, because the assumption of the utilized NB model that the features are normally distributed is not always correct. Furthermore, it can be observed that for increasing numbers of k, precision increases while recall decreases.

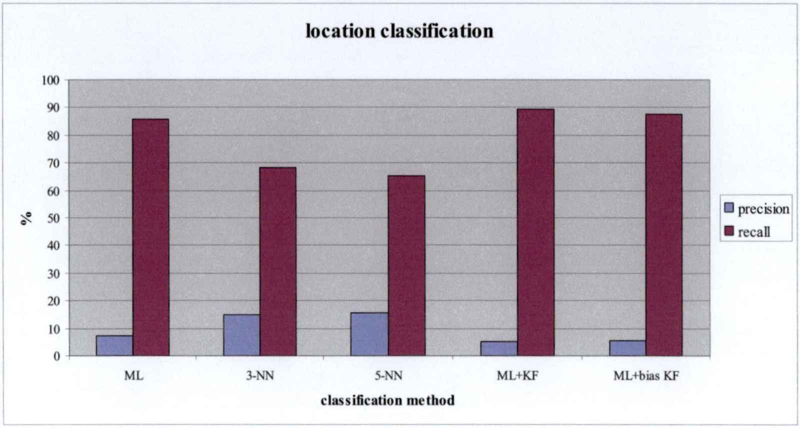

Figure 4.10: Location classification results of the maximum likelihood (ML) and the k-NN classifier (for k=3 and 5) in the PC scenario (recognition rates are averaged over all classes). Additionally, results of ML location classification applied to Kalman filtered data is shown for the 9-state (KF) and the 15-state (bias KF) EKF models.

4.5.3. Location Classification

Results from location classification based on VT measurements are depicted in figure 4.10. In the figure the performances of the ML method and two different k-NN classifiers are compared. The threshold value for the null class has been determined for ML classification by calculating the density values at 3 standard deviations around the normal distribution centers.

The comparison of the ML method and the k-NN classifiers shows that the k-NN classifiers yield better precision rates. However, the low recall rates of the k-NN classifiers also indicate that many locations have not been detected. Therefore, only the ML classification results are used in further experiments. In another experiment the VT position data and the IMU measurements has been fused with the 9-state and the 15-state EKF models before application of the ML location classification method. The result shows that both methods are able to increase the number of detected locations due to their capability to compensate short VT outages. On the other hand, they also increase the number of FP's. An explanation for this is that the KF increases the sampling rate,

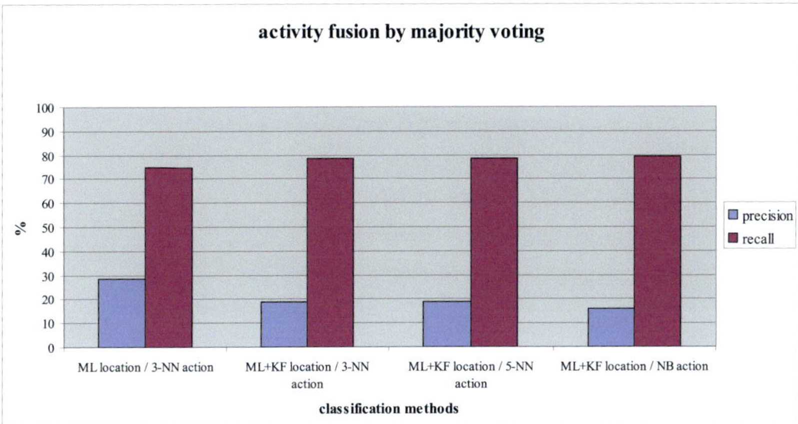

Figure 4.11: Comparison of activity fusion results by majority voting applied to different combinations of location and action classification methods.

which in turn leads to an increased number of false detections. Furthermore, besides several VT outages, comparatively fast rotational movements occur in this scenario. Some of these rotations are off the range of the angular rate sensors. This effect produces outliers in the KF results. Since reinitialization takes less time for the 9-state EKF than for the 15-state EKF, the 9-state EKF is producing less outliers in these situations and therefore yields a slightly better recognition performance. Because of that the 9-state EKF should be preferred to the 15-state EKF in this scenario.

4.5.4. Activity Fusion

Figure 4.11 shows a comparison of activity fusion results that have been obtained by applying the majority voting method to different combinations of location and action classification results. The first two results in the figure show that the KF also increases the number of TPs and FPs in the activity fusion result. The three results for which different action classification methods have been used (result 2-4) show again that the k-NN classifiers produce slightly reduced recall rates and better precision rates.

A class-wise account of the majority fusion result for the second action and

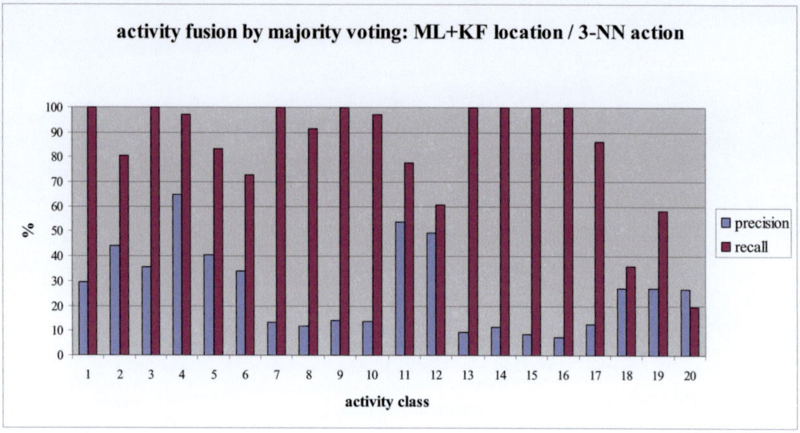

Figure 4.12: Class-wise activity recognition rates after activity fusion with majority voting.

location classifier combination is depicted in figure 4.12. It can be seen in the figure that half of the activities are recognized with recall rates of more than 95% and that only two activities are hard to detect with recall rates below 50%. On the other hand, the rates of false detections are relatively high for all classes (only two classes exist with a precision rate of more than 50%).
Compared to the result produced by location classification alone, many FP detections are cleared away after fusion of location and action recognition results. This can also be observed in the example of figure 4.8.

4.6. Summary

In this chapter, techniques for the online recognition of actions, locations and (low-level) activities from inertial sensor and position data has been presented. The information derived on this process level constitutes the decision input for task level activity recognition. Basically, all utilized methods are not restricted to any specific application and can also be used in other scenarios after recording of a new set of training data.
The proposed methods have been evaluated with test data from the PC scenario. In the tests, the nonparametric k NN action classification method has shown that it is able to yield better results than the naive Bayes approach.

However, for location classification the assumption of normally distributed positions proved to be more useful for achieving good recall rates than the nonparametric method. The utilization of Kalman filtered position data further improved recall rates at the cost of precision. The activity recognition result after the application of majority voting typically yields good detection rates, but still contains a lot of false alarms caused by confusions. However, because of the fact that low-level activity recognition results are used as input information for the recognition of task level activities, false rejections produce more costs than false detections. Therefore, in a scenario with complex worker behavior, low-level activity recognition constitutes a technique for obtaining input information for the recognition of task level worker activities with methods that are able to deal with uncertainty.

5. Task Level Activity Recognition

Task level activity recognition comprises the recognition of human behavior of the complexity of worker tasks. According to the solution concept in section 2.3, tasks can be modeled as sequences of low-level activities. This means that on the task recognition process level additional information in the form of context knowledge is incorporated in the activity recognition process. Besides the recognition of worker tasks, the utilization of context knowledge also allows to reexamine sequences of detected activities in order to produce an enhanced activity recognition result.

In the following, the methods that are utilized in this work for task level activity recognition are described and an evaluation of task level activity recognition results based on experiments with the PC scenario is presented. The techniques that are utilized in this chapter are represented by the state-based task recognition component in the human-machine interface overview in figure 2.2.

5.1. Introduction

In the view of our worker activity recognition concept, the recognition of worker tasks or subtasks constitutes a superordinate process level to low-level activity recognition. Thus, for the recognition of worker tasks, low-level activity information is processed on a higher level of abstraction by means of models that incorporate the interrelations between low-level activities. Since worker tasks in industrial environments are almost always defined as sequences of task steps (or activities), state-based task models are appropriate methods for the description of worker tasks.

A commonly used model for the description of human behavior is the finite state machine (FSM). An example for the utilization of a FSM as a formal

model for the description of human task performing processes for computerized manufacturing systems is given in [94]. In [101] a FSM based on inputs from processed wearable sensor signals is used to model a car production task. Another example is given by the work of [6], in which detections from vision-based office activity recognition are used as inputs in a state machine model for the recognition of human behavior in an office environment.
An important drawback of FSMs is that they assume that the input information is correct. This means that if detection results from low-level activity recognition are used as inputs they may not be affected with uncertainty. Therefore, an alternative approach that allows to incorporate uncertain detection results in high-level activity models are generative models. In [85] and [120] hidden Markov models (HMM) are used for the recognition of human behavior from probabilistic input information. Similarly, [70] uses an extension of a HMM to recognize high-level behaviors for vision-based office surveillance. In [121] household activities are recognized from RFID sensor signals and processed video data by means of probabilistic reasoning with a dynamic Bayesian network (DBN).
In this work statecharts are considered for comfortable task modelling in applications where it is possible to recognize low-level activities without any noise or uncertainty. Furthermore, since generative models are adequate techniques for the description of the sequential nature of worker tasks, a HMM and a DBN approach are compared for online task recognition with uncertain input information.

5.2. Task Modelling with Statecharts

Finite state machines are a common method for the deterministic description of worker tasks. However, as mentioned before, FSMs are based on the assumption that the input information is not affected with uncertainty. Therefore, the utilization of FSMs is limited to scenarios where reliable information about performed activities is available. Examples for this type of input information are accurate position tracking results, binary information (e.g. from binary switches or RFIDs) or internal data from utilized industrial tools or machines.

FSMs can formally be described by state diagrams that are based on tuples containing the input alphabet, the set of states, the start state, the state-transition function, such as the set of final states or the output function and the output alphabet. A drawback of conventional state diagrams is that the number of states increases with the number of possible parameter combinations and thus tends to become unmanageably large for many descriptions of behavior. To overcome this problem, an extended formalism called statecharts or Harel statecharts [35] has been introduced for describing complex behavior. Statecharts have become part of the Unified Modeling Language (UML) and extend state diagrams with hierarchies, concurrencies and communication features.
A statechart model is utilized for the recognition of worker tasks from reliable input information in real time experiments with the spot welding scenario in chapter 7.

5.3. Task Recognition with Hidden Markov Models

The HMM is a technique to model state-based processes, of which the current state is not known, but measurements that are related to the current state can be observed. Since HMMs are probabilistic models, they are able to deal with uncertainties or noise in measured observations. Originally HMMs have been utilized in the field of speech recognition. However, over the years they have successfully been applied to many other pattern recognition problems, such as activity recognition, gesture recognition or the analysis of peptide sequences.
In the following, a brief introduction into HMMs is given and our approach for the application of HMMs to the recognition of worker tasks is explained. For further reading on the theory of HMMs [79], [27] and [21] are recommended.

5.3.1. First Order Hidden Markov Models

For the first order HMM it is assumed that the propagation of the hidden state of an observed system follows a first order Markov chain. This means that at each step in time the state S_i of the system changes into a subsequent state S_j with the transition probability a_{ij}, which only depends on S_i and no preceding state earlier in time. The transition probabilities of a first order HMM with N hidden states can be expressed by the transition matrix:

$$A = \begin{pmatrix} a_{11} & a_{12} & \cdots & a_{1N} \\ a_{21} & a_{22} & & a_{2N} \\ \vdots & & \ddots & \\ a_{N1} & a_{N2} & & a_{NN} \end{pmatrix}, \tag{5.1}$$

in which the transition probabilities of each row have to be normalized in order to meet the conditions of a probability distribution:

$$\sum_{j=1}^{N} a_{ij} = 1; \forall i. \tag{5.2}$$

Furthermore, it is assumed that at each step in time a (discrete) HMM emits one out of M observations O_k that is related to the current state S_j with the probability b_{jk}. The observation probabilities are expressed by the observation matrix:

$$B = \begin{pmatrix} b_{11} & b_{12} & \cdots & b_{1M} \\ b_{21} & b_{22} & & b_{2M} \\ \vdots & & \ddots & \\ b_{N1} & b_{N2} & & b_{NM} \end{pmatrix}. \tag{5.3}$$

As in equation (5.2) the observation probabilities in each row of B have to be normalized according to:

$$\sum_{k=1}^{M} b_{jk} = 1; \forall j. \tag{5.4}$$

Thus given, a HMM is fully described by the notation

$$\lambda = (\vec{\pi}, A, B), \tag{5.5}$$

which contains the transition and the observation matrices A and B, such as the initial state probability distribution $\vec{\pi}$.

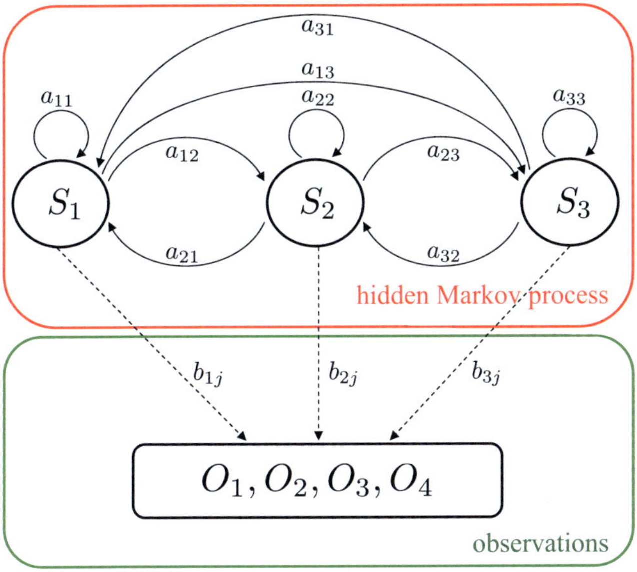

Figure 5.1: HMM Example with three hidden states S and four observations O.

Figure 5.1 depicts an example of a HMM, in which the hidden Markov process and the visible observations are indicated by the red and the green box.

5.3.2. Basic Hidden Markov Model Problems

When a HMM is applied to a pattern recognition problem, three different issues are of interest. These issues are also called the basic HMM problems.

Evaluation Problem

Given a sequence of observations O and a HMM λ, a common problem is to determine the probability $P(O|\lambda)$ that the observation sequence has been emitted by the model. The calculation of $P(O|\lambda)$ can be solved by means of the forward-backward algorithm. In HMM applications, $P(O|\lambda)$ is often utilized as a score value that indicates how well a sequence O and a model λ match.

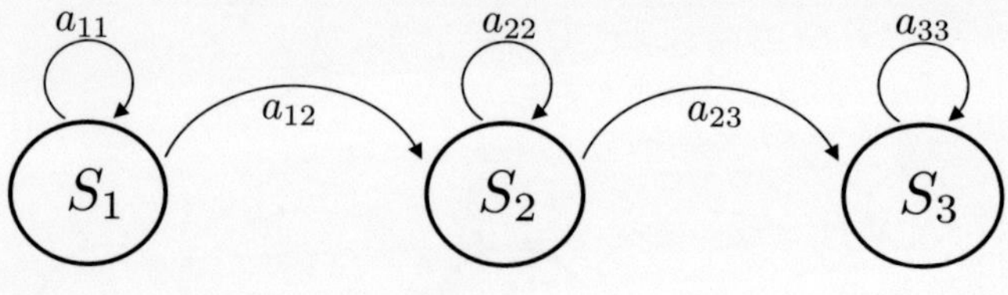

Figure 5.2: Left-to-right HMM example that allows only transitions to the current or the next state.

Decoding Problem

Besides the evaluation problem, another issue is to determine a state sequence S that corresponds to a given sequence of observations O, which has been produced by a known HMM λ. For finding the best fitting sequence S in an efficient way the Viterbi algorithm is used. Thus, the hidden state of the underlying Markov process can be estimated.

Learning Problem

Finally, as with other machine learning techniques a HMM has to be adapted to a recognition problem by parameter estimation. The learning problem constitutes the training of the HMM, that is to adapt the HMM parameters $\vec{\pi}$, A and B to a set of training data by maximization of $P(O|\lambda)$. The learning problem is solved with the Baum-Welch algorithm.

5.3.3. Hidden Markov Model Topologies

The HMM topology describes possible state-transitions of the underlying Markov process. Often the fully connected HMM (for an example of a fully connected HMM see figure 5.1) contains transitions that are unrealistic in practical applications. Therefore, possible transitions in a HMM should be restricted before training by setting unwanted transitions in the transition matrix to 0.

The left-to-right HMM is a HMM topology, in which no prior state can be reached from a current state. This property is used for instance to describe word models in speech recognition applications. In our HMM task recognition approach, tasks are modeled with left-to-right models, in which the states

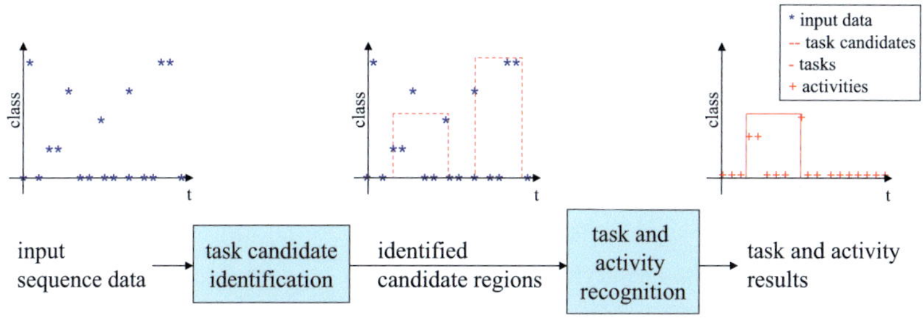

Figure 5.3: Online task and activity recognition with uncertain input data. In the first step task candidate regions are extracted from sequences of uncertain input data. Found candidates are examined in a second step in order to recognize tasks and task related activities.

correspond to worker activities. The used left-to-right HMMs are restricted to allow only transitions to the current or the next state, that is:

$$a_{ij} = 0; \quad if\ j < i\ or\ j > i+1. \tag{5.6}$$

An example of such a restricted left-to-right model is given in figure 5.2.

5.3.4. Online Hidden Markov Model

For the online recognition of tasks and activities in sequences of uncertain low-level activity recognition data with HMMs a two step approach is utilized that first identifies task candidate regions and then validates found candidates in a second step (see also figure 5.3 for an illustration). Each worker task is represented in this process by a single HMM that contains worker activities as HMM states. Since abrupt transitions from one activity to another are unrealistic in real-world applications, null class states are inserted between activity states and at the start and the end of each task model. As observations the activity class results from majority fusion are used, which are related to the respective task HMM.

In the task candidate identification step in figure 5.3 the HMMs of all tasks are applied in parallel to sequences of detected activities in order to identify regions of task candidates. Since the recognition process is performed online, extraction windows of a fixed length are used to extract observation sequences for each HMM. The length of the extraction windows is the maximal assumed duration of the respective task. By analyzing the Viterbi path task candidate regions can be identified with their exact start and end points in the extraction window.

In the task and activity recognition step in figure 5.3 the HMM is applied again to identified candidate regions in order to determine the likelihood $P(O|\lambda)$ only for the observation sequence part of a candidate region. The task candidate is classified as a detection of a performed task if a threshold comparison of the likelihood yields a positive result and if the activity states in the Viterbi path have been generated to a certain percentage from non-null observations. The Viterbi path of a correctly detected task then contains the HMM activity results.

5.4. Task Recognition with Dynamic Bayesian Networks

A limitation of HMMs is that the hidden state of a modeled system is directly linked to a set of observations. Because of this limitation, only fused activities can be used as observations. Otherwise, the set of observations would become large if observations are generated by using all possible combinations of action classification and location classification results. However, a large set of observations is not desirable for the HMM, since it would cause overfitting problems, if the set of training data is limited. Therefore, activity fusion is essential for the practical application of the HMM in order to keep the number of observations reasonably small. On the other hand, activity fusion on the low-level activity recognition level means that the information content is reduced without being able to utilize context information.

A method to overcome this problem is to model the task recognition process with a DBN, which is able to incorporate distributed input information, such as action classification and location classification results. Because of the fact that the HMM can be seen as a special case of a DBN, the utilization of DBNs instead of HMMs does not mean a change to the basic idea of the probabilistic task recognition approach.

In the following sections, the DBN theory is briefly explained and an online task recognition procedure with DBNs is introduced.

5.4.1. Bayesian Networks

In this section a brief introduction into Bayesian networks (BN) is given for providing a basic understanding of DBNs. More detailed information on BNs and DBNs can be found in textbooks or publications such as [46], [33] or [86].

A BN is a directed acyclic graph (DAG), which contains random variables as nodes and dependencies between random variables as edges. In a BN, a node is called parent of another node, denoted as child, if there exists a directed edge from the parent node to the child node. It is assumed in a conventional BN model that for each child node the conditional probabilities of the node (i.e. the random variable) are given depending on its parent nodes (e.g. in the form of probability tables). For nodes in the network that have no parents the prior probabilities are given instead.

D-Separation

A method to determine if two variables in a network are independent is the d-separation rule. The d-separation rule depends on the evidence about variables that lie on paths between the variables for which independence is to be determined, whereas evidence about a random variable means that the state of the variable is known. The d-separation rule further depends on how variables are connected. Figure 5.4 shows an example of a BN that contains 3 different types of possible connections between variables: the serial connection, the diverging connection and the converging connection. A connection between A and D is called serial connection. In this case, A and D are d-separated, if the state of B is known, since the state of A then has no more influence on D. A diverging connection can be described as a parent node with two or more child nodes. An example for this connection type is the connection between D, E and F. E and F are d-separated if the state of D is known. Finally, the connection between B, C and D is denoted as converging connection. In this

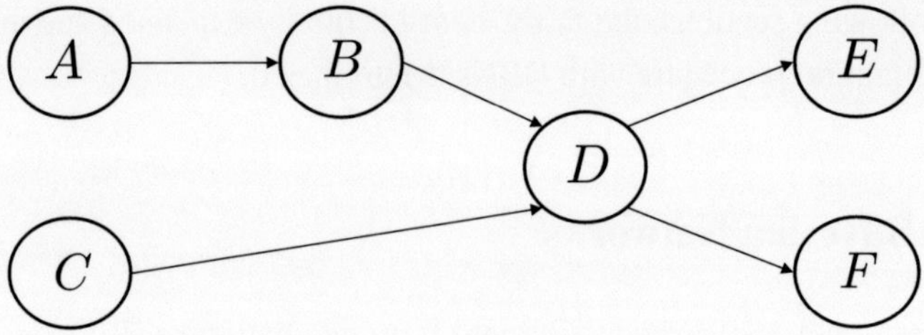

Figure 5.4: Example of a Bayesian network with the 6 variables A, B, C, D, E and F.

connection the variable B is independent of C if D is not known. Thus, B and C are d-separated if the state of D (and the state of potential descendants of D) is not known.
In summary, the d-separation rule can be formulated as [46]: Two variables A and B in a BN are d-separated if there is an intermediate variable V on the path between A and B and either V is known and the connection is serial or diverging or V and its descendants are not known and the connection is diverging.

Parameter Learning

Given a set of training data and the structure of a BN the parameters of the BN (i.e. the probability tables) can be determined. A common technique to learn the probability tables of a BN is to use a maximum likelihood estimation method. By means of this method the probabilities of the states of a variable are calculated depending on the states of its parents by regarding the ratios of state occurrences in the training data instances. The following equation gives an example for the calculation of the conditional probabilities of the variable D in figure 5.4:

$$P\left(D=d_i|B=b_j,C=c_k\right)=\frac{N\left(D=d_i,B=b_j,C=c_k\right)}{N\left(B=b_j,C=c_k\right)}. \tag{5.7}$$

In equation (5.7), the probability $P\left(D=d_i|B=b_j,C=c_k\right)$ of D being in state d_i in dependence of B being in state b_j and C being in state c_k is determined by calculating the ratio of the numbers N of training data instances, in which the respective states of the involved variables occur. Application of equation (5.7) to all combinations of states of D, B and C yields all conditional probabilities of the probability table. A drawback of the maximum likelihood estimation method is that conditional probabilities may be determined as 0 if the set of training data is small or not representative. A method to avoid this is to incorporate initial probabilities in the estimation procedure, e.g. in the form of a uniform distribution.

Inference

In applications of BN it is often of interest to determine the probability of the state of a variable x_i given the states of other variables, denoted as evidence E. This probability is given by the following equation :

$$P(x_i|E) = \frac{P(x_i, E)}{P(E)}. \tag{5.8}$$

For obtaining $P(x_i|E)$ in equation (5.8), $P(x_i, E)$ has to be determined first. Assuming that we can calculate the probability over all variables (also called the universe) in a BN $P(U)$, we are able to determine the probability of a subset of all variables by marginalization. For instance, $P(E,A)$ in the BN of figure 5.4 can be determined as

$$P(E,A) = \sum_{B,C,D,F} P(U) = \sum_{B,C,D,F} P(A,B,C,D,E,F). \tag{5.9}$$

Thus, the probability of E given the evidence $A = a_i$ would then be given by $P(E, A = a_i)$.
The probability over all variables of a BN with N nodes can be determined by means of the chain rule

$$P(U) = \prod_{n=1}^{N} P(X_n|pa(X_n)), \tag{5.10}$$

where $pa(X_n)$ denotes the parents of X_n.
A drawback of the above mentioned procedure is that the calculation of $P(U)$ leads to a probability table that grows exponentially with the number of variables in the network. Therefore, it is advisable to avoid extreme computing costs by utilization of an efficient algorithm (such as the junction tree algorithm) for determining probabilities in large BNs.

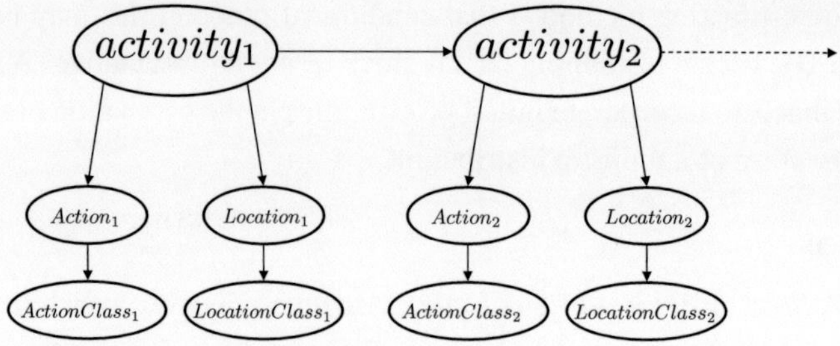

Figure 5.5: DBN for modelling a task as a progress of activities. The activities are nodes in the BNs that model the dependencies between activities, actions and locations, such as classification results.

5.4.2. Dynamic Bayesian Networks

DBNs constitute an extension of the idea of BNs by incorporation of a temporal aspect. DBNs are composed of BNs that exist at discrete points in time, called time slices. Usually it is assumed that the same BN is used at every time slice. The BNs of the different time slices are connected by edges that are directed towards future time slices. This leads to a large network that allows to include evidences at certain points in time and to calculate the state of variables at different points in time. Thus, a DBN offers the possibility to model the temporal behavior of a system of variables with probabilistic dependencies.

Figure 5.5 shows the DBN, which is used for modeling a task as a progress of activities depending on actions and locations. In the figure two time slices of the task recognition DBN are depicted. Each time slice consists of a BN, in which the dependencies between activities, actions and locations are modeled. The nodes *ActionClass* and *LocationClass* represent the results from action and location classification. For these variables, evidences are available if the task DBN is applied to sequences of classified actions and locations. Since it must not be the case that classified actions and locations are correct, they are not directly connected with the activity nodes, but through the mediating variables *action* and *location*.

5.4.3. Online Dynamic Bayesian Network

The DBN online recognition procedure utilized in this work is similar to the online HMM process. As shown in figure 5.3 task candidates are identified in a first step and performed tasks and activities are recognized in a second one. However, since the DBN is able to process information from distributed sources, action and location results are used as DBN evidences instead of the result from activity fusion. Another difference to the online HMM process is that a single DBN is used for the recognition of all tasks. This can be achieved by modelling the activity probability table in a way that is similar to the transition matrix of a task HMM and by including all tasks and activities. After the identification of the exact start and end points of task candidates in the first step, the DBN is applied again to found candidate regions. A candidate region is then classified as executed task if the respective activities of the task could be recognized to a minimum percentage.

5.5. Experiments and Results

The task level activity recognition approach has been evaluated with the dataset that has been recorded in the PC scenario experiment described in section 4.5.1. The evaluation has been conducted with the same evaluation method (threefold cross-validation and TP/FP counting method for activity recognition results) as explained in section 4.5.1. Since the low-level activity recognition results are affected with uncertainty in all PC scenario experiments, the statechart method is not adequate for application to the PC scenario dataset and a probabilistic method is required. Therefore, the evaluation contains a comparison of the HMM and the DBN approach (results from experiments with the statechart method and spot welding scenario are shown in chapter 7).

5.5.1. Activity Recognition Results

In figure 5.6 an example for activity recognition results produced with the HMM and the DBN is given. Since the HMM uses the activity fusion result

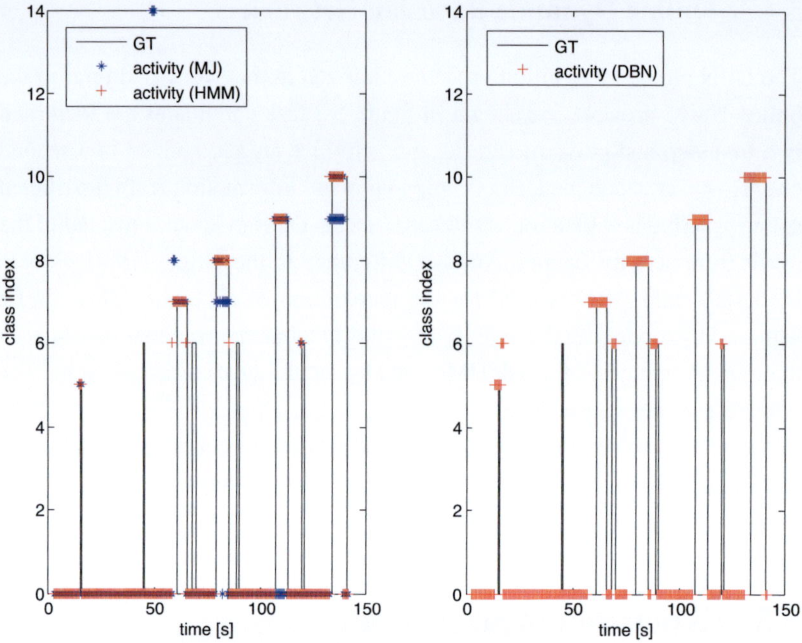

Figure 5.6: Result example for activity recognition with the HMM (left) and the DBN (right). Additionally, the majority voting result (MJ) is shown in the left plot as a basis for comparison. The test sequence, which has been used in this example, is of task class number 2.

from majority voting as input, these results are additionally shown in the left plot. It can be seen in this plot that the HMM is able to eliminate several false detections of the majority fusion result by means of the incorporated context knowledge. However, missing the detections after activity fusion cannot be recovered. Therefore, some detections of class 6 for which no activity result has been available do not overlap with the GT (they have been inserted arbitrarily by the HMM). The result plot for the DBN on the right of the figure shows a similar result. However, since the DBN uses the unprocessed classification results as inputs, it has more input information for decision making. This explains that the activities of class six have been recognized more accurately with the DBN.

A statistical evaluation (averaged over all activities and participants) of ex-

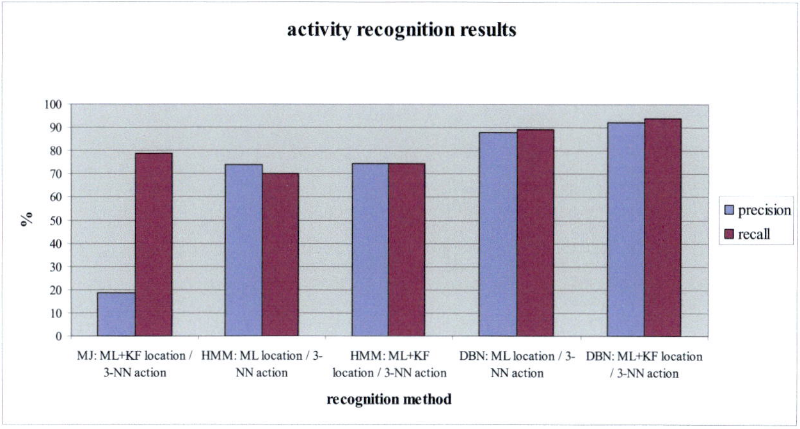

Figure 5.7: Comparison of activity recognition results. All tests are based on the detections from maximum likelihood location classification (with and without KF) and 3-NN action classification. For the HMM experiments location and action classification results have been fused by majority voting.

periments with all test sequences of the PC scenario dataset is shown in figure 5.7. For a better comparison, the majority voting result from section 4.5.4 that has been created with detections from maximum likelihood location classification (with Kalman filtered positions) and 3-NN action classification is shown on the left of the figure. The comparison with the majority voting result shows that the recall rate is slightly reduced if the HMM is applied, but it can also be observed that the precision rate is significantly improved in the HMM result. It can further be observed that the DBN is able to significantly improve the recognition rates. Besides that it can again be seen that the KF improves the recognition rates.

5.5.2. Task Recognition Results

Figure 5.8 shows an example of HMM and the DBN results that have been generated with a sequence of input data containing all tasks of the PC scenario. In the figure the complete activity recognition result is shown on the top and the task recognition result is shown on the bottom. It can be observed that the HMM and the DBN show a similar behavior as in the experiment of

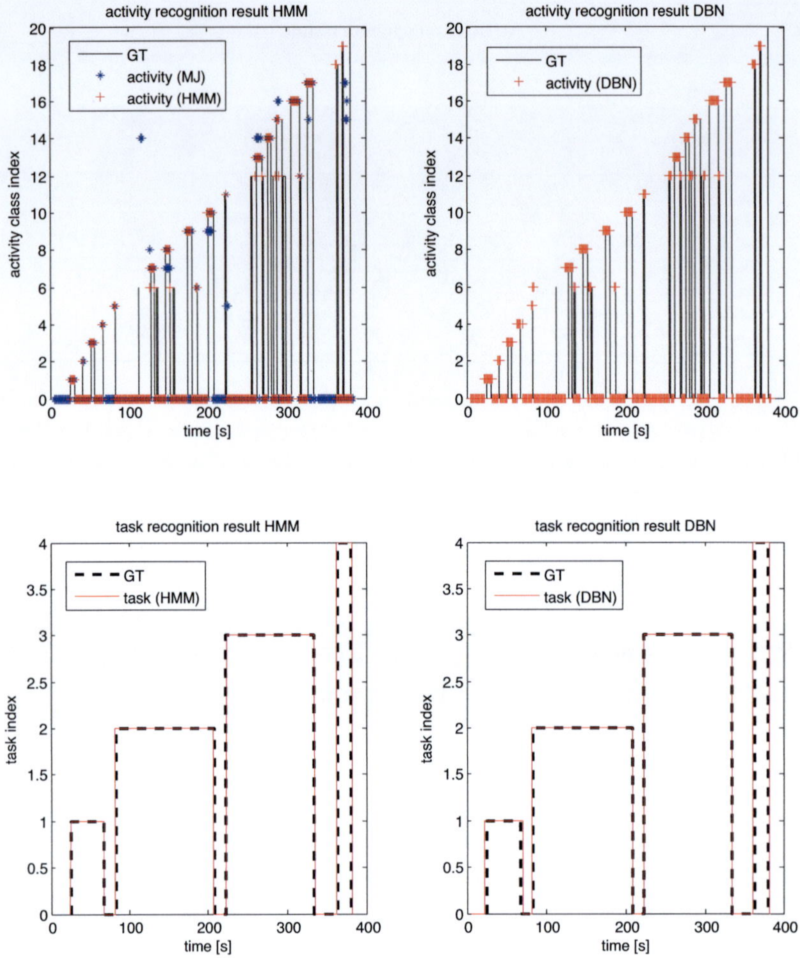

Figure 5.8: Result example for activity (top) and task (bottom) recognition with the HMM (left) and the DBN (right). The processed sequence contains all tasks of the PC scenario.

figure 5.6, in which the results of the subsequence of task number two in figure 5.8 had been depicted. Having a look at the task recognition result at the bottom of figure 5.8, we can see that the recognized tasks match accurately with the GT for both methods.

Figure 5.9 shows a statistical evaluation (averaged over all tasks and partic-

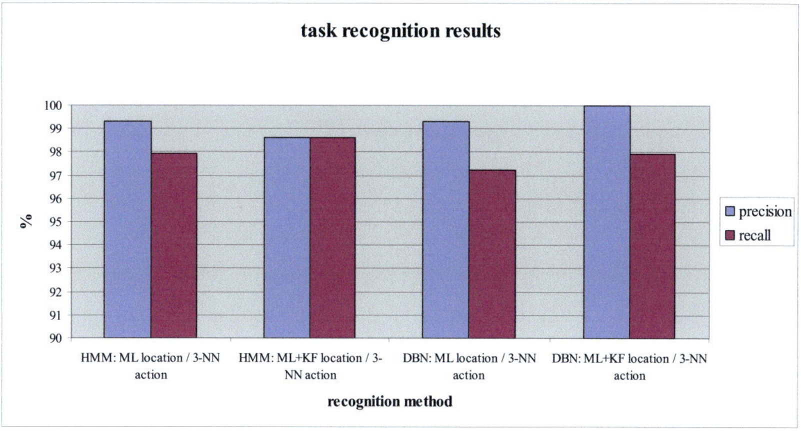

Figure 5.9: Comparison of online task recognition results based on sequences containing all tasks. All tests are based on the detections from maximum likelihood location classification (with and without KF) and 3-NN action classification. For the HMM experiments location and action classification results have been fused by majority voting.

ipants) of the task recognition results for sequences of the PC dataset containing all 4 tasks. In the evaluation, detected tasks that overlap with the GT have been counted as TPs and otherwise FPs. The result in the figure shows that all results yield high recognition rates and differ not significantly if the HMM or the DBN is applied. Therefore it can be said that both, the HMM and the DBN yield high task recognition results. However, it has to be noted that the DBN tends to produce more accurate results in the sense that start and endpoints of detected tasks match better with the GT than HMM results in some cases. The reason for this is the better activity recognition capability of the DBN.

Another issue, which should be regarded for practical applications is the consumption of computing time. The complexity of the DBN causes long processing times compared to the HMM. For instance, the DBN needed 110s on our test PC for processing a sequence of 389s length in reality. In contrast to that, the online HMM approach needed less than 0.6s for processing the whole sequence.

5.6. Summary

In this chapter three different methods for the recognition of worker behavior on the task level have been presented. By means of the task level activity recognition methods, activities and tasks can be recognized from low-level activities (or action and location information) by making use of context knowledge.

For the recognition of tasks from reliable input data, which means that the data is not affected with uncertainty, a statechart model has been proposed. As an alternative, task level activity recognition with uncertain input data is possible by using probabilistic state-based models. The proposed online HMM and online DBN approaches have both proven their usefulness by yielding high task recognition rates in experiments with the PC scenario with precision and recall rates that are far above 95%. Both methods are also able to yield significantly better activity recognition results than the majority fusion method, which does not incorporate context knowledge. The best activity recognition results have been produced by utilizing the DBN (with location input data from Kalman filtered positions). However, because of the complexity of the DBN comparably much computing time is necessary to process data with this model.

Finally it can be said from our point of view that the incorporation of context knowledge is essential for solving worker activity recognition problems that are of a complexity of the chosen PC scenario.

6. Gesture Recognition System

In this chapter, our system for online human gesture recognition is presented. The gesture recognition system is represented by the gesture recognition and the (time series representation) feature extraction components of the human-machine interface in figure 2.2. The term online refers to the detection of gestures in streams of sensor data (i.e from the IMU in figure 2.2). Thus, our goal is not to classify isolated gesture templates, but to spot and classify gestures, which are present in unsegmented time series of undefined length.
The purpose of the gesture recognition system is to provide the possibility of comfortable communication with a superordinate framework to the worker. This comprises the recognition of a small number (in the order of 5-10) of hand gestures for interactive communication. Because performing gestures means to mimic predefined movements, repeated gestures show higher spatio-temporal similarities than task related human behavior, which concentrates on fulfilling certain task goals rather than on executing a task in a certain manner. Thus, spatio-temporal similarities within classes of gestures are an important issue to be considered for the gesture recognition system. The developed system may not only be used for human gesture recognition, but also for the recognition of human behavior with spatio-temporal features in general (which makes it a flexible component in particular).
In the following sections an online template matching method and an optimization method for finding representative gesture templates, called prototypes, are described. The presented approaches and parts of the results can also be found in our publications [39] and [36].

6.1. Introduction

The first design factor of a gesture recognition system is the choice of appropriate sensors. Since a lot of work has been done in the field of gesture recognition with the focus on a variety of applications, different types of sensors have been utilized.

Data gloves are expensive, but offer the observation of many degrees of freedom of a human hand. Therefore, they have primarily been used for sign language recognition [40, 58, 60].

Camera based systems have been utilized for recognizing sign language [96] or gestures [59, 118]. However, vision-based systems suffer from external influences and disturbances, such as bad illumination and occlusions. Furthermore, they require complex image processing methods in order to detect hands or other parts of the body [114].

In recent years, inertial sensors (especially accelerometers) have been gathering more and more interest in different gesture recognition applications, such as human computer interaction [11, 90], or human machine interaction [106]. Advantages of using inertial sensors are their insensitiveness to disturbing external influences and reasonable costs. For these reasons and because of the fact that they are already an existing component of our system, accelerometers are used for our gesture recognition system.

Another important development issue in the gesture recognition system design is the choice of the technique with which the sensor signals are processed. In contrast to human activity recognition, where rather generalizing approaches are used, gestures can be described as human behavior with strong spatio-temporal similarities within each class of gesture signals. As an example, figure 6.1 shows three instances of a gesture of the same class, which have been recorded with a triaxial accelerometer. From the figure it can be observed that all instances have a very characteristic shape (disregarding spatial and temporal variations). This explains why in current research usually techniques are utilized, which model gestures as time series focusing on spatio-temporal information in the signal.

Because of the similarity of problems in the research fields of speech recognition and gesture recognition, time series models from the field of speech

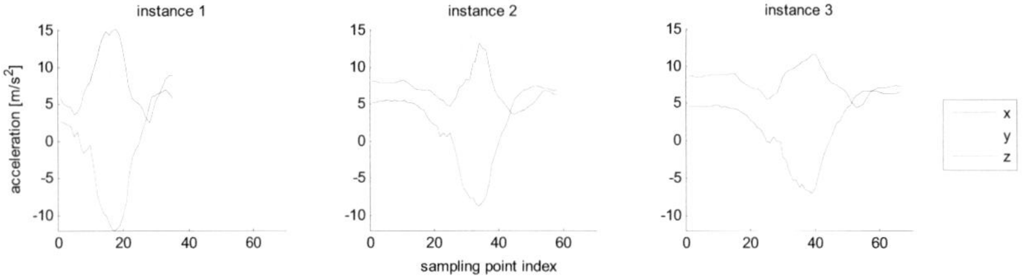

Figure 6.1: Gesture signals recorded with a triaxial accelerometer: The three instances of the gesture resemble each other in shape, but show spatial and temporal variabilities.

recognition are often used for recognizing gestures. Commonly used types of time series models can be roughly divided into probabilistic methods and template matching techniques. Among probabilistic models Hidden Markov Models (HMM's) are very popular for gesture recognition [18, 53, 90]. Template matching approaches consist of a time series representation and a distance measure. The results of an experimental comparison in [20] show that common representations differ not significantly in performance. For this reason, we use median-filtered accelerometer signals, which is similar to the Piecewise Aggregate Approximation (PAA) representation. However, it is also shown that the choice of elastic distance measures improve classification performance for small training sets. The term "elastic" means in this context that the distance measure is able to cope with temporal variations in measured signals. This is a very important issue, since gestures are usually not repeated at exactly the same speed (this can also be observed from figure 6.1). A commonly used elastic time series distance measure for gesture recognition applications is dynamic time warping (DTW) [17, 54, 61].

Because of user-dependent variabilities in the execution of gestures, user-dependent systems yield better performance than user-independent systems. But we also have to consider that for our application a gesture recognition system is needed that offers a certain flexibility by means of the ability of easy adaptation to new conditions. Therefore it is important to keep the size of the training set low in order to avoid exhaustive recording and to stay flexible. Since HMM's require larger sets of training data and DTW is able to compete with HMM's in gesture recognition [54, 61], we decided to utilize DTW as distance measure.

A key problem in the application of time series recognition techniques is that they require representative templates, also called prototypes or motifs. A lot of work has been published, in which different approaches for the selection or generation of prototypes for time series template matching in different applications such as handwriting recognition [80], speech recognition [117], visual pattern recognition [122] or general time series analysis applications [65, 68] are described. However, many published approaches are restricted in their applicability to distance measures with particular properties (i.e. metrics or measures, which fulfill the triangle inequality [26, 69]) or are adapted to specific techniques [31]. Furthermore, motif finding techniques usually focus on templates, which yield a good representation of a certain class [95]. However, separability of prototypes from other classes of templates is often not considered. In this work a novel approach for finding time series prototypes is presented, which is not restricted to specific distance measures and aims at maximal class separability.

6.2. Dynamic Time Warping for Gesture Recognition

The basic principle of time series template matching is to determine how well a given test template T matches with a prototype template C in order to find out whether T belongs to the class of C.
DTW is a template matching technique, which is able to cope with temporal variations in signals. Because of the fact that spoken language varies in speed, the method has originally been applied to the field of speech recognition [88]. Since similar effects occur when human gestures have to be recognized the method can be found in gesture recognition applications as well [54].

6.2.1. Dynamic Time Warping Algorithm

The DTW algorithm is a template matching method that measures the similarity between time series templates with temporal variations by employing a dynamic programming technique.

Let us assume that we are given two time series templates:

$$C = (c_1 \dots c_N) \tag{6.1}$$

and

$$T = (t_1 \dots t_M). \tag{6.2}$$

C and T may resemble each other in general appearance, but are not temporally aligned to each other with respect to their sampling points c_n and t_m and have different lengths N and M. Because of that a direct pointwise comparison (e.g. using Euclidean distance) would possibly not yield a good match. The basic idea of DTW is to calculate the matching score from best-fitting pairs of sampling points (c_n, t_m). Therefore, a so-called warping path

$$W = [(wc(1), wt(1)) \dots (wc(L), wt(L))] \tag{6.3}$$

has to be found, in which the element pairs $wc(l)$ and $wt(l)$ contain the indices of the sampling point pairs from which pointwise distances are calculated. The warping path result of the DTW algorithm can then be understood as the solution to the optimization problem that minimizes the sum over all pairwise distances $d(c_n, t_m)$. Thus, the DTW distance (or DTW score) is given by:

$$DTW(C,T) = \min_W \left(\sum_{l=1}^{L} d(c_{wc(l)}, t_{wt(l)}) \right). \tag{6.4}$$

Because of the fact that there is no standard implementation of the DTW algorithm, different implementations can be found in literature. Typical implementation considerations and algorithm variants, which may be used in practical applications, are listed in the following.

Point distance measure

As distance measure $d(\cdot)$ for the sampling point pairs (c_n, t_m) in equation (6.4) usually the Euclidean distance is used, however using other (positive) distance measure or norm is possible as well.
Because of the penalizing effect on outliers, the squared Euclidean distance measure is used in our DTW implementation. The Euclidean distance between two points $\vec{x}$ and $\vec{y}$ in a F-dimensional feature space is defined as:

$$d(\vec{x}, \vec{y}) = \sum_{f=1}^{F} (x_f - y_f)^2. \tag{6.5}$$

Path constraints

Since allowing arbitrary combinations of sampling point pairs would lead to warping paths with no practical relevance, some restrictions have to be made depending on the application. Therefore, W has to be restricted to realistic solutions, which is typically achieved by using the following constraints [24]:

- Monotonicity: Warping path indices are monotonically increasing (going back in time is not allowed): $wc(l) \geq wc(l-1)$ and $wt(l) \geq wt(l-1)$.
- Continuity: Limitation of the increment between successive elements (if limited to 1 index increment no jumps are allowed): $wc(l) - wc(l-1) \leq 1$ and $wt(l) - wt(l-1) \leq 1$.
- Boundaries: The warping path of two templates C and T begins at their first elements and ends at their last elements, respectively: $wc(1) = 1$ and $wt(1) = 1$ such as $wc(L) = N$ and $wt(L) = M$.

Besides these commonly used constraints additional variations for calculating DTW paths are existing.
In [88] and [78] several path weights are proposed for normalizing warping paths with respect to their length.
Furthermore, the search space for warping paths can be restricted by utilizing band restrictions such as the Sakoe-Chiba band [88] or the Itakura parallelogram [43]. Because band restrictions limit possible combinations of sampling point pairs (c_n, t_m) to realistic subsets, extremely shifted match solutions of C and T are avoided and computing time is reduced.
Our DTW implementation uses no special path constraint variations, besides the common path constraints and a Sakoe-Chiba band.

Variable start and endpoints

In some applications (such as word recognition [78] or gesture recognition [54]) it may occur that time series C and T contain non-relevant information (e.g. noise) at their starts and ends. In this case the warping path can not be constrained to start at the first and end at the last template elements (boundary constraint). Thus, W may start at template elements $wc(1) = S_c$, $wt(1) = S_t$ and end at template elements $wc(L) = E_c$, $wt(L) = E_t$ with $1 \leq S_c < E_c \leq N$

and $1 \leq S_t < E_t \leq M$.
Utilizing variable end points is necessary for our application, as it will be explained later in section 6.2.3.

Another important issue to mention is that since DTW is basically a distance measure for time series, prototype dependent thresholds have to be determined if DTW is used for classification of time series. The threshold can be determined empirically from a set of training data (e.g. by evaluating the typical distances from the prototype to templates of the same class).

6.2.2. Range Normalization

In some gesture recognition applications, measured sensor signals do not only vary in their temporal appearance, but show also spatial variations. These spatial variations result either from sensor characteristics (e.g. biased signals) or from gestures that are repeated in a different way and cause varying amplitudes in the quantity to be measured. The latter occurs especially with gesture signals that are recorded with accelerometers (which can be seen in figure 6.1). Varying amplitudes in gesture signals from accelerometers are caused by two effects. The first is that accelerometer readings are influenced by gravity and that this influence depends on the attitude of the body part with which the gesture is performed. And, secondly, acceleration amplitudes differ within a class of gestures, because humans usually do not repeat gestures at exactly the same speed. These effects lead to templates, which appear biased or scaled to each other.
Typical implementations of the DTW algorithm allow to cope with temporal variations in time series signals, however variations in the spatial signal content can not be handled. In order to allow comparisons of biased or scaled templates, time series are normalized to the same range of values in our DTW implementation. A one-dimensional time series $A = (a_1 \ldots a_N)$ can be normalized to the $[0,1]$ range via:

$$A_{norm} = \frac{A - \min(A)}{\max(A) - \min(A)}. \tag{6.6}$$

The normalization is applied to both templates C and T before calculating the DTW distance $DTW(C,T)$. However, in order to avoid unrealistic scalings, the ratio between the scaling factors

$$r_{RN\_scal} = \frac{\max(C) - \min(C)}{\max(T) - \min(T)} \tag{6.7}$$

is limited in our approach. Without the scaling limit, templates of a certain class could be confused with random noise. Maximal scaling factors are determined by maximal scaling ratios in the training data set.

In case of multidimensional time series comparisons, range normalization is applied to every time series component of multidimensional templates independently.

6.2.3. Online Gesture Recognition

In an online gesture recognition system, gestures have to be recognized from a stream of sensor data. This makes it necessary that templates are extracted from the data stream, before template matching can be applied. In some applications segmentation of the data stream is possible by motion detection. However, this is not possible in an industrial environment, since a worker is not only moving for performing interactive gestures. Furthermore, the problem of variable start and endpoints has to be considered.

In our online implementation, raw data measured from accelerometers is first processed with a median filter in order to obtain data in an appropriate (subsampled and noise-filtered) time series representation. Then, templates in form of signal parts are extracted from the filtered data stream by means of a sliding window in order to determine the DTW distance to a prototype for threshold classification.

Since we utilize a prototype optimization method (see section 6.3), it can be assumed that the prototypes do not contain non-relevant information. Because of that the problem of variable start and endpoints can only affect the time series templates, which are extracted from the data stream. Furthermore, considering that templates are extracted with a sliding window, variable start points can be neglected as well, if window spacings are narrow enough. Therefore, we only have to consider that templates in the extraction window

have variable end points, depending on the length of the gesture. This can be achieved by allowing the warping path to end in the region $E1 \leq wt(L) \leq E2$. The parameters $E1$ and $E2$ can be determined for each prototype from the minimal and maximal length of templates of the relevant class in training set. The online DTW algorithm works then as described as follows. A sliding window spanning a length of $E2$ points from a start sampling point n to $n+E2$ is slided with a spacing of one or more points over the stream of filtered data. After that, the DTW distance from the prototype C to the template T extracted by the window is calculated. By means of the sliding window and by incorporating the variable end region between $n+E1$ and $n+E2$, matching gestures of variable length can be found in the data stream.
Usually, multiple gestures are to be recognized in one datastream. This makes it necessary to execute the online DTW procedure in parallel for each used prototype. As exemplified in figure 6.2, for each gesture class one sliding window of a prototype depending length $E2$ is used. Gestures are then detected, if the match score in a window falls below the threshold of the respective prototype.

6.2.4. Ambiguity Resolution

The presented sliding window method for online gesture recognition can lead to overlapping detections. These ambiguities can occur either as overlapping detections of one specific prototype or as overlapping detections of different prototypes in the multiclass case. Depending on the spacing of the sliding window, overlaps of one specific prototype usually occur around the area of a gesture match. These kinds of ambiguities can easily be resolved by taking the detection with the lowest DTW distance and eliminating all overlaps. A problem arises when ambiguities are caused by detections from different prototypes, because due to prototype depending thresholds the DTW results are not directly comparable. In order to solve this problem we introduce a relative score value by modelling DTW distances from a class prototype C_i to other templates of class i as probability distributions. Assuming an underlying normal distribution, DTW distances can be normalized for obtaining a

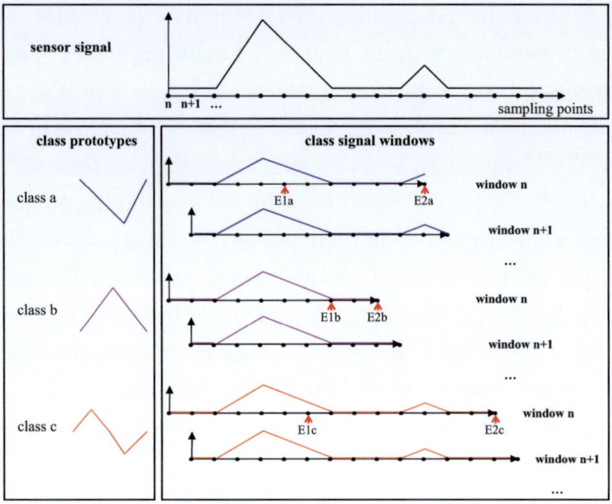

Figure 6.2: Online gesture recognition with multiple time series prototypes. Templates are extracted in parallel from an input data stream for distance calculations to each class prototype. Depending on the prototype, sliding windows with prototype depending lengths ($E2$) are utilized.

relative DTW distance $DTW_{relat}(C_i, T)$ according to:

$$DTW_{relat}(C_i, T) = \frac{DTW(C_i, T) - \mu_i}{\sigma_i}. \tag{6.8}$$

Mean value μ_i and standard deviation σ_i in equation (6.8) can be determined from a set of training templates for each prototype.
Since the relative DTW distance offers a measure to directly compare detections from different prototypes, ambiguities from overlapping detections of different classes can now be resolved by taking the detection with the lowest relative DTW distance.

6.3. Prototype Optimization

The general idea behind template matching is that a class of templates is represented by one or more prototype templates. Because of that, the choice of appropriate prototypes is essential for the performance of a template matching method.

It has to be noted here that we focus on finding one representative prototype for each class of gestures in order to keep the amount of templates manageable. Thus, representation of a class by multiple prototypes is not considered in our application.

As mentioned in section 6.1, there has already some work been published about the selection or generation of time series prototypes from a set of training templates. Typically used methods for DTW prototype selection are to chose the template with the best recognition performance or the smallest distance to all other templates within a class of templates [54] or to use clustering methods [78]. Instead of prototype selection, new DTW templates can be generated by temporal alignment and averaging of all templates of a particular class ([1], [54]). Alternatively, other approaches from the field of time series motif mining can be utilized for choosing DTW prototypes, if their preconditions are met by the DTW distance measure.
The above mentioned methods for DTW prototype selection and generation such as many other approaches for finding time series prototypes have in common that representative prototypes are chosen by their distance to templates of the same class. In contrast to that, our understanding of a representative prototype is based on the separability of time series classes. Therefore, besides distances of a prototype to templates of the own class, distances of a prototype to templates of other (known) classes are additionally incorporated in our measure of representativeness. In the following, the distance of a prototype to a template of its own class is named intraclass distance and the distance of a prototype to a template of a different class is named interclass distance.

Figure 6.3 depicts a schematic example, in which DTW distances from three prototypes of class C to templates of classes C and T) are shown as point distributions. If we now have a look at the the intraclass distances (marked as blue circles) it appears that on average prototype 1 has smaller interclass distances than the other prototypes. However, for obtaining a measure of class separability, we also have to consider the interclass distances (marked as red x). In the figure it appears that the class separability between intraclass

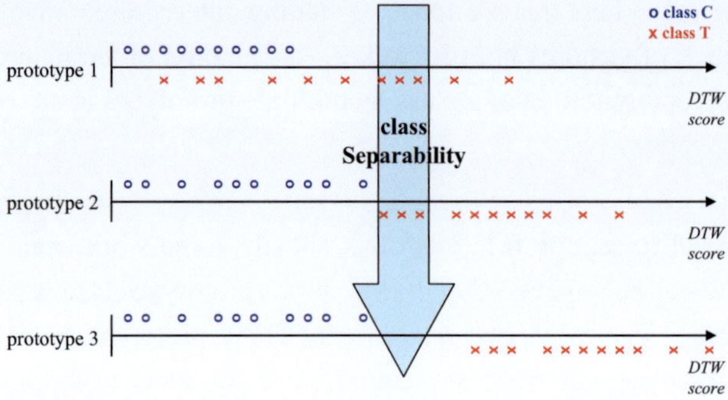

Figure 6.3: Distance distributions of DTW score values. The distance distributions of the prototypes to templates of classes C and T differ depending on the prototype. Prototype 3 achieves the best separation between the distributions.

and interclass distance distributions increases from prototype 1 to 3. Thus, prototype 3 would be a the best choice according to our definition of a representative prototype.

In [124] an approach for finding prototypes in subseries of time series sequences has been presented, which is based on information gain. However, although somewhat similar to our idea, the approach in this work utilizes symmetric and non-elastic distance measures.

Besides the definition of a measure for the representativeness of a prototype, another important issue is how to select or generate prototypes from a set of training templates. In common gesture recognition applications, gesture templates are usually cut out from recorded gesture data either by visual inspection or by making use of labels, which have been assigned during recording. This means that start and end of a gesture are defined manually in order to obtain a set of templates. However, because of the fact that manual labelling usually contains inaccuracies, it can not be guaranteed that cut out templates contain exactly the signal content of a gesture. Furthermore, due to variations in parts of gesture signals, the representative content may be found in the subseries of a gesture template. For these reasons, a representative prototype must not necessarily be the whole time series of a (cut out) template, but is often found in template subseries. A practical solution to this problem is to extract templates from manually labeled data with an additional tolerance in a first step in order to avoid potential loss of relevant information. Then, representative prototypes can be found in the subseries of these templates by using a search or optimization method.

Expressed in formal terms, the problem for choosing a representative prototype C_{opt} for a certain class of templates C can be described as the following optimization problem:
Assuming that we are given a set of time series templates containing I instances of a particular class

$$C_i = (c_{i,1} \dots c_{i,N_i}) \quad ; i = 1 \dots I \tag{6.9}$$

and J instances of other known template classes

$$T_j = (t_{j,1} \dots t_{j,M_j}) \quad ; j = 1 \dots J, \tag{6.10}$$

our target is now to find a prototype

$$C_{opt} = (c_{i,\alpha} \dots c_{i,\beta}) \quad ; 1 \leq \alpha < \beta \quad \alpha < \beta \leq N_i \tag{6.11}$$

in the subseries of templates C_i, which optimally represents class C in the sense that it yields the best separation between the distance distributions of classes C and T. Our definition of class separability is based on the intraclass and the interclass time series distances and can be described in form of target functions, which will be explained in more detail in section 6.3.1. Since relevant gesture signal contents are present in subseries of templates of classes C and T, flexible start and endpoints have to be regarded in all distance calculations. Template subseries, which yield the best fit to the prototype C_{opt} are denoted as C_i' and T_j'. The parameters α and β span a search space, in which the solution of the optimization problem C_{opt} can be found by means of a search procedure or an optimization method (see section 6.3.2).
Although DTW is used as a distance measure for gesture time series in this work, the presented prototype optimization approach is not limited to this particular distance measure, but can be used with any other time series distance measure.

6.3.1. Target Functions

The formulation of choosing representative prototypes as an optimization problem makes it necessary to define a target function as a measure of class representativeness. As mentioned in the previous section, our definition of representativeness is based on the class separability, that is the maximization of the spread between intraclass and interclass distance distributions of template classes as depicted in figure 6.3. Since there exists no general target function model for this problem, we defined four target functions of different complexity, which are explained in the following. These target functions measure the class separability of a (candidate) prototype C_{opt} based on a training set of templates of the prototype class C and other classes T_k. It has to be noted that all defined target functions measure class separability by considering absolute class distances. Absolute class distances are more expressive than relative measures such as entropy or information gain (e.g. used in [124]).
In order to obtain a decision criterion for classification, a prototype depending distance threshold τ has to be determined for each found prototype. The determination of this threshold depends on the utilized target function model and is explained in this section, as well.

Minimal Interclass to Maximal Intraclass Distance (min_max)

A straightforward approach for measuring class separability is to take the difference between the minimal interclass distance and the maximal intraclass distance. At the top of figure 6.4 an illustration for this measure is given, which is described by the min_max target function:

$$f_{min\_max}(C_{opt}) = \min_{j,k}(DTW(C_{opt}, T'_{j,k})) - \max_{i}\left(DTW(C_{opt}, C'_i)\right). \quad (6.12)$$

Although simple to use and conservative in the sense that this measure incorporates the worst case distance difference of a template dataset, a disadvantage of the min_max target function is that it is prone to outliers.
The classification threshold τ_{min_max} for a found prototype can be determined according to figure 6.4 by calculating the average between minimal interclass

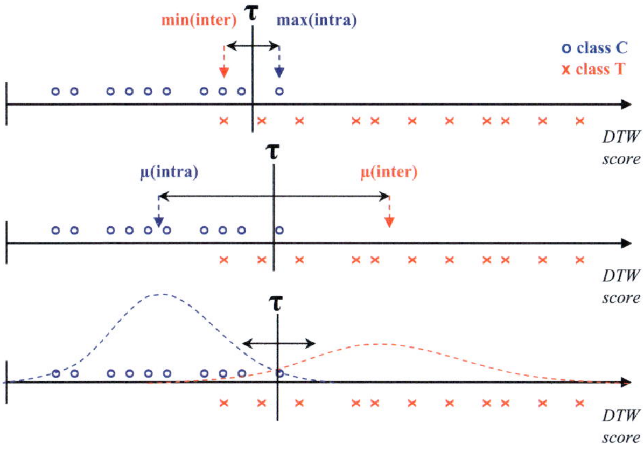

Figure 6.4: Different target function models applied to a distribution of DTW score values produced by a prototype of class C (from top to bottom): Minimal interclass to maximal intraclass distance, center point distance and distribution-based distances (Kullback-Leibler divergence and error function integral model).

distance and maximal intraclass distance:

$$\tau_{min\_max}(C_{opt}) = \max_i \left(DTW(C_{opt}, C'_i)\right) + \frac{\min_{j,k}(DTW(C_{opt}, T'_{j,k})) - \max_i (DTW(C_{opt}, C'_i))}{2}. \tag{6.13}$$

Center Point Distance (CP_dist)

Another approach for a target function that measures class separability is to calculate the distance between the center points of intraclass and interclass distance distributions, as indicated in the middle of figure 6.4. Given a prototype C_{opt} and the template training set, the center point of the intraclass distance distribution is calculated as

$$\mu_{cc} = \frac{1}{I} \sum_{i=1}^{I} DTW(C_{opt}, C'_i) \tag{6.14}$$

and the center points of the interclass distance distributions are calculated as

$$\mu_{ct,k} = \frac{1}{J} \sum_{j=1}^{J} DTW(C_{opt}, T'_{j,k}). \tag{6.15}$$

Based on the center points given in equations (6.14) and (6.15) the CP_dist target function can then be formulated according to:

$$f_{CP\_dist}(C_{opt}) = \frac{1}{K}\sum_{k=1}^{K} log(\mu_{ct,k}) - log(\mu_{cc}). \tag{6.16}$$

The logarithm in equation (6.16) attenuates potential overweighting influences, which is necessary in case if there are classes among the K classes of templates T_k with comparable high distances. Compared to the min_max target function, the CP_dist target function is a more generalizing measure and robust to outliers, because of the averaging effect of the mean value calculation.

Classification thresholds τ_{CP_dist} for found prototypes are calculated analogous to equation (6.13):

$$\tau_{CP\_dist}(C_{opt}) = \mu_{cc} + \frac{\min_k(\mu_{ct,k}) - \mu_{cc}}{2}. \tag{6.17}$$

Kullback-Leibler Divergence (KL_div)

The min_max and the CP_dist target functions are based on information from interclass and intraclass distance distributions, resulting from a prototype C_{opt}. However, both measures do not include the whole distribution information, because not all distribution characteristics are considered.

A measure of the difference between two probability distributions $p_1(\vec{x})$ and $p_2(\vec{x})$ that includes all distribution characteristics is the Kullback-Leibler divergence [56]:

$$D_{KL\ 1,2}(p_1(\vec{x}), p_2(\vec{x})) = \int_{-\infty}^{\infty} p_1(\vec{x})\, ln\frac{p_1(\vec{x})}{p_2(\vec{x})} d\vec{x}. \tag{6.18}$$

Assuming that the intraclass and interclass distances are normally distributed, we define the target function based on the Kullback-Leibler divergence (KL_div) as follows (see appendix A for the derivation of the symmetric divergence formula):

$$\begin{aligned} f_{KL\_div}(C_{opt}) &= \frac{1}{K}\sum_{k=1}^{K} sgn(\mu_{ct,k} - \mu_{cc}) \cdot log\left(\frac{1}{2}\left(\frac{\sigma_{ct,k}^2}{\sigma_{cc}^2} + \frac{\sigma_{cc}^2}{\sigma_{ct,k}^2} - 2\right)\right. \\ &\left.+\frac{1}{2}(\mu_{cc} - \mu_{ct,k})^2\left(\frac{1}{\sigma_{cc}^2} + \frac{1}{\sigma_{ct,k}^2}\right)\right), \end{aligned} \tag{6.19}$$

in which the intraclass and interclass standard deviations σ_{cc} and $\sigma_{ct,k}$ are given by:

$$\sigma_{cc} = \frac{1}{I-1}\sqrt{\sum_{i=1}^{I}\left(DTW(C_{opt},C_i')-\mu_{cc}\right)^2} \tag{6.20}$$

and

$$\sigma_{ct,k} = \frac{1}{J-1}\sqrt{\sum_{j=1}^{J}\left(DTW(C_{opt},T_{j,k}')-\mu_{ct,k}\right)^2}. \tag{6.21}$$

Similarly to equation (6.16), results from outlying classes are attenuated in equation (6.19) by means of taking the logarithm.
Assuming that the time series distances are class-wise normally distributed, an appropriate threshold for separating two classes can be determined by calculating the intersection point between the center points of their distributions. This threshold, which minimizes the classification error, is illustrated at the bottom of figure 6.4. Because of the fact that two normal distributions have two intersection points s_1 and s_2 (unless they have the same standard deviation) with $s_1 < s_2$, the following case differentiation has to be made to determine the relevant intersection point:

$$\tau_k = \begin{cases} s_1, & \text{if } \sigma_{cc} > \sigma_{ct,k} \\ s_2, & \text{if } \sigma_{cc} < \sigma_{ct,k} \end{cases}, \tag{6.22}$$

Since the threshold τ_k in equation (6.22) separates class C from T_k, k threshold results exist if $k > 1$. Therefore, we decide for the smallest threshold value in the multiclass case:

$$\tau_{min}(C_{opt}) = \min_k(\tau_k). \tag{6.23}$$

It has to be noted that for determining the prototype threshold τ_{min}, only thresholds τ_k are considered in equation (6.23) for which $\mu_{ct,k} \leq \mu_{cc}$.

Error Function Integral (erf_int)

A direct approach to measure the separability of normal distributions is based on the integral over the distribution functions. In the example in figure 6.4, this can be achieved by calculating the integral of the intraclass distribution over $[-\infty,\tau]$ and the integral of the intraclass distribution over $[\tau,\infty]$. Theoretically, the result is then inversely proportional to the threshold classifica-

tion error, which can be expected if the preconditions are fulfilled and the training set is representative.
Since the Kullback-Leibler divergence measures solely the dissimilarity between probability densities, this target function is a more exact measure for separability. However, for the integration over normal distribution function, the Gauss error function [2] has to be utilized, which increases the complexity of the target function:

$$erf(x) = \frac{2}{\sqrt{\pi}} \int_0^x e^{-t^2} dt. \tag{6.24}$$

The cumulative normal distribution function $F(x)$ can then be expressed by [47]:

$$F(x) = \frac{1}{2} + \frac{1}{2} erf\left(\frac{x-\mu}{\sqrt{2}\sigma}\right). \tag{6.25}$$

If we now first determine the classification threshold according to equation (6.23), the erf_int target function score is then calculated with the distribution characteristics from equations (6.14), (6.15), (6.20) and (6.21) after the following formula:

$$\begin{aligned} f_{erf\_int}(C_{opt}) &= F_{cc}(\tau_{min}) + \frac{1}{N}\sum_{n=1}^{N}\left(1 - F_{ct,n}(\tau_{min})\right) \\ &= \frac{1}{2} + \frac{1}{2} erf\left(\frac{\tau_{min} - \mu_{cc}}{\sqrt{2}\sigma_{cc}}\right) \\ &\quad + \frac{1}{N}\sum_{n=1}^{N}\left(\frac{1}{2} - \frac{1}{2} erf\left(\frac{\tau_{min} - \mu_{ct,n}}{\sqrt{2}\sigma_{ct,n}}\right)\right). \end{aligned} \tag{6.26}$$

Equation (6.26) contains the intraclass distribution integral and the averaged interclass distributions integrals over the area of correct classifications. In order to additionally penalize bad separations, interclass distribution integral results are set to 0 if $\mu_{ct} \leq \mu_{cc}$ in equation (6.26).

6.3.2. Optimization Methods

After having defined several target functions for measuring the representativeness of a prototype, a method has to be chosen in order to find representative prototypes C_{opt} in the subseries of a training template set. This makes

it necessary to define the parameters of a search space and to decide for an appropriate optimization method for performing a parameter search.
Based on a set of templates of class C and the prototype definition in equation (6.11), the search space is spanned up by the template index i and the parameters α and β, which represent start and end of the subseries of a template C_i. Since templates in a training set have different lengths and are not temporarily aligned to each other, there is no logical link between their sampling points. This makes it inevitable to perform the search independently for each template C_i on the set of their subseries.
Thus, for each template C_i a representative prototype candidate $C_{opt,i}$ has to be found in the search space spanned up by the parameters α and β:

$$f_{target}\left(C_{opt,i}\right) = f_{target}\left(\alpha,\beta\right) \longrightarrow max. \tag{6.27}$$

Based on the target function scores $f_{target}\left(C_{opt,i}\right)$, found prototype candidates can then be compared in order to chose the most representative prototype

$$C_{opt} = \arg\min_{i}\left(f_{target}\left(C_{opt,i}\right)\right). \tag{6.28}$$

In the following, two different methods are presented, which have been utilized for performing the parameter search.

Brute Force Search

The most straightforward method to perform a search in a (discrete) search space is a brute force search. Brute force search means to search the space spanned up by α and β by testing every point (α,β) with the target function in order to find the maximum.
The advantage of the brute force search is that it is guaranteed to find the optimal solution, which makes it appropriate for using it as benchmark. However, brute force search is inefficient, because of its trivial strategy it consumes the maximal possible computing time. This makes it not applicable for large sets of data.

Evolution Strategy

Since a brute force search can be time consuming, a more efficient search method has to be applied in case of large datasets. The first step for choosing an appropriate search method is to determine the category of the optimization problem. Having a closer look at the target functions in section 6.3.1 shows us that we are dealing with a nonlinear optimization problem. Furthermore, it can not be excluded that there exists more than one maximum in the search space, which means that our optimization problem is not local. Because of theses facts, we decided to use an evolution strategy (ES) to tackle this complex optimization problem.

The evolution strategy [92] is a heuristic optimization technique, which belongs to the class of evolutionary algorithms. It has been developed by Rechenberg [82] in the 1960s and is based on a natural understanding of the optimization problem. The general approach of the ES is to have a population of μ arbitrary points in the search space, which produces a number λ of randomly generated descendants over some iterations. According to Darwin's law "survival of the fittest", only the best solutions survive over time and are able to generate descendants. The resulting effect is that the fitness (i.e. the target scores) of individuals in the population increases over time until a satisfying solution is found.

Figure 6.5 shows a flow chart of the ES algorithm, which contains the following process steps and mechanisms:

Initialization

The algorithm is initialized with a population of μ individuals, which are generated arbitrarily. In order to make sure that the ES yields a solution that is at least as good as the original template C_i, our initial population always contains this template as an individual. Furthermore, an initial step length is set in this step. The step length is a parameter, which controls the variation of the parameters in the mutation step.

Mutation

In the mutation step, the population is augmented by creation of a number of λ new individuals. Being descendants of the μ old individuals, new individuals are created by random variation of parameters of their parents. The parameter variation is normally distributed with zero mean and a standard deviation of the current step length σ. Adapted to our problem, parameter

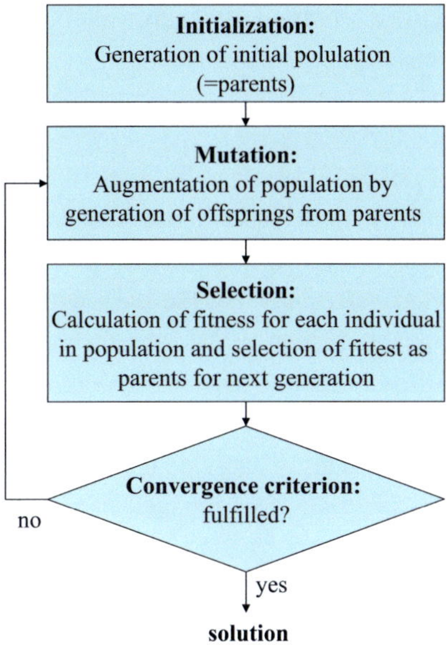

Figure 6.5: Flow chart of the evolution strategy algorithm.

variation is described as follows:

$$\begin{pmatrix} \alpha_{descendant} \\ \beta_{descendant} \end{pmatrix} = \begin{pmatrix} \alpha_{parent} \\ \beta_{parent} \end{pmatrix} + \begin{pmatrix} z_\alpha \\ z_\beta \end{pmatrix}, \tag{6.29}$$

with $z_\alpha \sim \mathscr{N}\left(0, \sigma_\alpha^2\right)$ and $z_\beta \sim \mathscr{N}\left(0, \sigma_\beta^2\right)$.

Step length control

The step length control is an adaptive mechanism for controlling the parameter variation in the mutation step. Although parameter variation happens randomly, the average variation can be controlled by means of the current step length σ. Step length control is important, because due to regional differences of target function values in the search space, larger (e.g. in flat regions) or smaller steps (e.g. in steep regions) are effective.

In [92] the $1/5$ success rule is proposed, which is applied after a certain number of iterations. By means of this rule, the step length is decreased or increased if less or more than $1/5$ of the mutations are successful. A successful mutation means in this context that a descendant has a better target score value than its parent. The proposed rate with which the step length σ should be increased or decreased is a factor of 0.85.

Selection

After the creation of new individuals in the mutation step, the population is reduced to a number of μ individuals in the selection step in order to keep the population constant. Since the goal of optimization is to find the global extremum, the selected individuals are the ones with the best target scores (i.e. the fittest ones). Depending on the search strategy, different rules are existing for the selection of individuals. In the comma strategy ((μ,λ) with $\lambda > \mu$), the μ surviving individuals are selected out of the λ descendants, which leads to the effect that an individual only lasts for one generation. In contrast to this, all individuals of the current population are considered for selection in the plus strategy $(\mu + \lambda)$. Since we want to make sure that already found solutions may persist, a plus strategy is used in this work.

Convergence criterion

At the end of an iteration, found solutions are checked if they fulfill a convergence criterion. If the criterion is fulfilled the algorithm terminates and the individual with the best target score out of the current μ individuals is taken as solution to the optimization problem. Otherwise, the current μ individuals form the next generation and the process is repeated with the mutation step. Checking for convergence must not necessarily be done for each generation, but also after a certain number of iterations.
There are different rules existing for the decision about convergence. Convergence criteria can be to test the step length or the distance of the parameters $(\Delta\alpha, \Delta\beta)$ from iteration to iteration in order to check if they fall below certain thresholds. Also, the change of the target function ($\Delta f(C_{opt})$) can be taken as measure for convergence. In order to limit the processing time and to have a simple rule for convergence (which does not require to determine thresholds depending on target functions), we decided to use a fixed number of iterations as abort criterion.

It has to be noted that since the ES is a heuristic optimization method, solutions are in principle not reproducible. Furthermore, found solutions may be not the exact minimum or maximum, but are usually close to an extremum. On the other hand there are a lot of important advantages over other optimization methods, which are the reasons why we decided to use the ES.
Most important advantages are that the ES is able to tackle nonlinear optimization problems and that the problem of getting stuck in a local extremum is unlikely. Other advantages are that the method is time efficient and to some extent generic, which means that there is no detailed knowledge of the problem required.

6.4. Gesture Recognition Process

In practical applications, the presented gesture recognition methods are applied according to the scheme depicted in figure 6.6. The process is split up into two parts, an offline training phase and a test or online gesture recognition phase.

6.4.1. Training phase

In the offline process part, representative prototypes are found and DTW parameters are determined. Based on a set of training data, which contains labeled time series sequences (in our application recorded with accelerometers), the training process starts with the extraction of templates belonging to different classes of gestures. After that, the templates are subsampled or filtered in order to obtain an appropriate time series representation. With this set of training templates, one representative prototype for each class is found by prototype optimization. Furthermore, DTW parameters are determined for each prototype, which are the classification threshold τ, the end region parameters $E1$ and $E2$, such as the maximal scaling factors for range normalization.
Excepting data recording and labelling, the training process runs fully automatically.

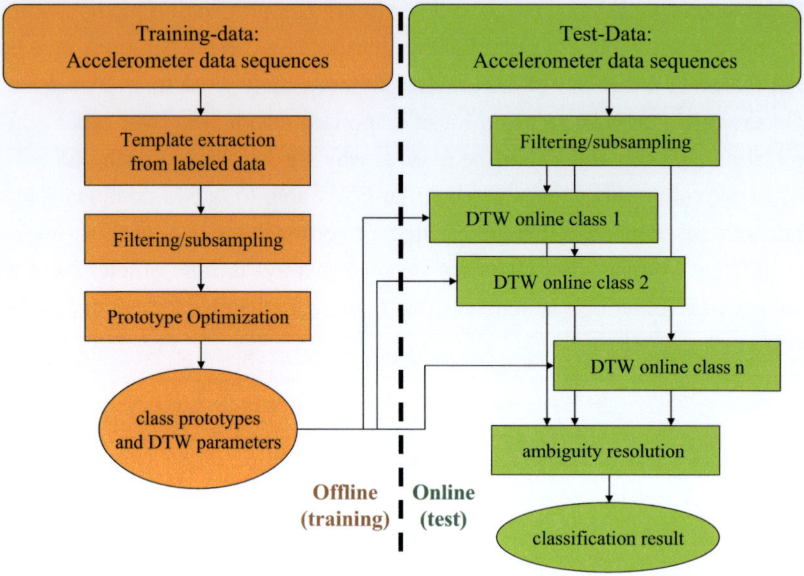

Figure 6.6: Scheme of the gesture recognition process. Class prototypes are extracted offline from training data in the training phase. Gesture recognition is applied in the online test phase to sequences of input data and is based on distance calculations to the class prototypes.

6.4.2. Online gesture recognition phase

The second part of the process constitutes the actual online gesture recognition process. After filtering of the incoming stream of sensor data by utilizing the same technique as in the training phase, the online DTW algorithm is executed in parallel in order to obtain the distances of the prototypes to segments of the stream as described in section 6.2. The gesture recognition result is then obtained by threshold classification of the calculated distances and eventual application of ambiguity resolution in case of the occurrence of ambiguous results. Online gesture recognition results exist in form of segments in the data stream, which match to the respective gesture template.

6.5. Experiments and Results

In order to validate the presented gesture recognition system, several tests have been performed with recorded gesture data. In this section, an outline of the test conditions is given and results from tests with online gesture recognition are presented.
It has to be noted that all tests have been performed with user-dependent data, since we have decided to use user-dependent gesture recognition in our application in order to have an adapted system with reliable performance rates. However, the fact that we utilize a template matching method keeps the required amount of training data in reasonable dimensions.

6.5.1. Dataset and Testing Procedure

Since gesture recognition is applied in this work in order to detect hand gestures for user interaction with a superordinate framework, a gesture dataset has been recorded with the sensor bracelet from section 3.4.2.
The gesture dataset as well as the testing and evaluation procedure are explained in the following.

Gesture dataset

For recording the test dataset, 9 different hand gestures have been defined as possible commands for user interaction. An instruction on how the gestures should be performed, which was given to test subjects, is depicted in figure 6.7 a). All gestures have been performed with a single hand while accelerations have been recorded with the sensor bracelet.

The test dataset contains data of 7 participants (5 male, 2 female) in form of sequences, in which the 9 gestures are performed after another. With each person, 15 gesture sequences have been recorded and start and endpoints of gestures have been labeled manually. Thus, time series segments, in which gestures occur, are marked with the label data in the gesture sequences. By means of the manually assigned labels isolated templates have been cut out automatically from sequence data for training. Furthermore, the label data serves as ground truth for evaluation of online gesture recognition. However,

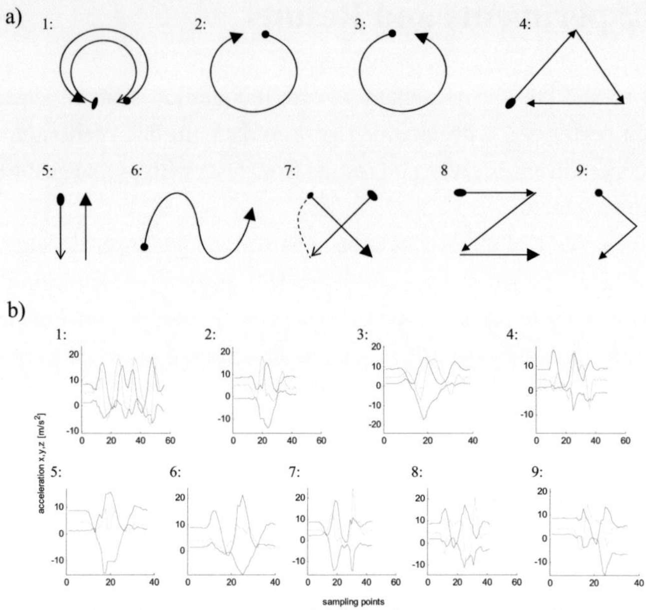

Figure 6.7: Gestures utilized in DTW tests: a) Gesture descriptions (arm movements). b) Accelerometer signals of the respective gestures.

it has to be considered that the occurrence timing inaccuracies in the label data is unavoidable. In total, the gesture dataset consists of 105 gesture sequences (each containing 9 gestures) from which 945 instances of isolated gesture templates can be extracted.

The sequences of raw accelerometer data have been recorded with the 3 orthogonal accelerometer channels of the ADIS16350 IMU at a sampling rate of 407 Hz (an example for the isolated templates of all gestures is given in figure 6.7 b)). In all tests a median filter has been applied to the raw sensor data for filtering and subsampling. The used median filter had a window width of 16 sampling points and a window spacing of 8 sampling points.

Testing and evaluation procedures

In all tests for evaluating prototype optimization and gesture recognition performances, the online DTW algorithm has been utilized. Prototype depending DTW parameters (i.e. end region parameters $E1$ and $E3$ such as maximal

scaling factors for range normalization) have been determined in each test by means of the particular training data. The window spacings in the online DTW algorithm were set to 1 sampling point.
The tests have been conducted in form of a (user-dependent) threefold cross validation (CV), which allows to use the whole dataset for testing without having overlapping sets of training and test data in a single test run. This means that in a single test run, 10 isolated template instances have been used for the optimization of DTW prototypes (according to the training phase on the left of figure 6.6) of each class of gestures. Testing (i.e. online DTW processing) has then been done with each prototype and the 5 sequence instances that do not contain the templates, which have been used in the training. These single test runs were conducted three times with the data of each participant for processing all 15 sequence instances.
As a basis for evaluation, true positives and false positives have been counted in tests in order to calculate the statistical measures precision and recall (which have been defined in equations (4.12) and (4.13)). We define true and false positives in online time series recognition experiments according to:

- True positives (TP) are detected segments, which overlap with the ground truth (GT) of the respective class. Multiple detections overlapping with one GT segment are considered as one TP and are not treated specifically.
- False positives (FP) are detected segments, which do not overlap with any GT segment or only with the GT segment of the wrong class. Analogous to multiple TP detections, FP segments that overlap with already counted FP's are not treated specifically as well.

As an example, figure 6.8 a) shows the detection result after application of the online DTW algorithm (using a prototype of the gesture class that is marked in the sequence with the GT) to the accelerometer data shown in figure 6.8 b). Detections of the DTW algorithm in figure 6.8 a) occur when the DTW score value (blue line) falls below the DTW threshold of the prototype (red line). By incorporating the length of detected DTW templates, result segments are marked with the magenta colored boxes.

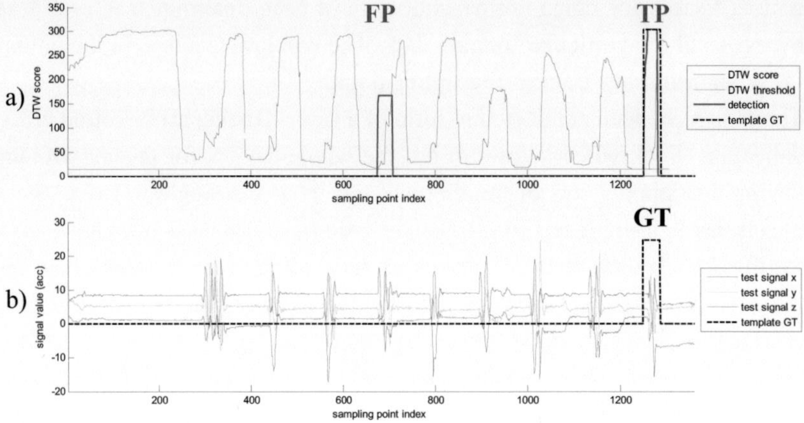

Figure 6.8: Online DTW applied to a gesture sequence (utilizing a prototype of class number 9): a) Online DTW score value, GT (black dashed line) and detections (one true positive overlapping with the GT and one false detection). b) Accelerometer signals (in m/s^2) with GT.

6.5.2. Prototype Optimization Results

The prototype optimization approach has been tested and evaluated with the gesture dataset and testing procedure from the previous section. Conducted tests concentrate on two issues. The first issue is a comparison of the BFS and the ES search strategies with respect to efficiency and applicability. Secondly, performance of optimization with the different target functions, i.e. their influence on the classification result has been tested and evaluated.

In the optimization tests a 5+5 ES has been used, which has been terminated after 15 iterations in order to have a deterministic processing time limit (these parameters have turned out to be practicable in prior experiments). These parameters have been determined experimentally by checking if the target function values show a converging behavior after the 15 iterations in performed tests. Initial ES populations contained the original template and 4 subseries for which the parameters α and β have been chosen randomly.

For all prototype optimization tests, online DTW with range normalization and no ambiguity resolution was used.

Comparison of BFS and ES

For comparison of the applicability of BFS and the ES optimization methods to prototype optimization, processing time and optimization performance tests have been conducted. All tests have been performed on a test PC with an Intel Core 2 Duo E8400 CPU and 2GB DDR2 RAM. Both optimization methods have been implemented in C++ and the test software has been run under Windows XP.
Because of the fact that BFS optimization is very time-consuming, tests have only been performed with the min_max and the erf_int target functions for optimizing templates of the randomly chosen classes 1 and 9 (utilizing the templates of CV sets 1 and 2, respectively). Since solutions found with ES optimization are non-deterministic, ES tests have been performed 10 times.
Figure 6.9 shows a comparison of BFS and average ES processing times measured in the tests. As expected, the BFS takes much more computing time than the ES (e.g. about 50h for the BFS and less than one hour for the ES in the min_max target function test with class 1 and CV set 1). Furthermore, it can be observed that the differences between minimal and maximal processing times of the 10 ES runs, which are shown by the error bars, are negligible small. The processing time differences between min_max and erf_int target functions appear to be minor, as well.

In figure 6.10, the progress of the target function values after each ES iteration is illustrated for each test run. At each step the maximal function result of all templates belonging to the optimized class and all μ individuals is displayed, which constitutes the optimization solution. Since the BFS optimization method searches the whole target space for the global target function maximum and the initial population of the ES contains the original templates, target function values of ES solutions range between the original result (green line) and the BFS maximum (red line).
It can be seen in the figure that in most of the test runs the ES optimization result shows a converging behavior after round about 10 iterations. This indicates that the choice of 15 ES iterations is sufficient for this problem. In almost all test runs, the ES result is close to the BFS result, which shows that the ES is able to find satisfying solutions.

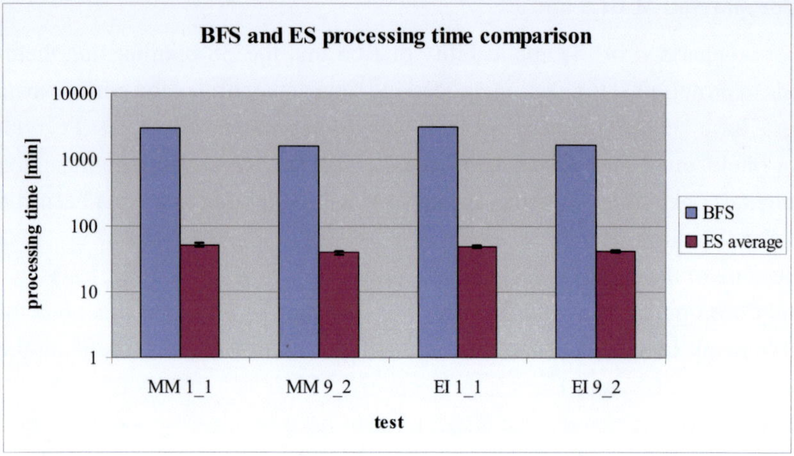

Figure 6.9: BFS and ES processing times. ES results are shown as averaged results over 10 runs (the error bars indicate the minimal and maximal processing time). Tests have been conducted with min_max (MM) and erf_int (EI) target functions for classes 1 and 9 (utilizing cross validation sets 1 and 2, respectively)

Target function performance

Figure 6.11 shows the results averaged over all classes of a comparison test with ES optimization and all target functions from section 6.3.1. Since templates obtained from manually assigned labels are potentially inaccurate, a tolerance of 500 sampling points (which is equal to 1.25 s) has been added to start and endpoints of the labels for template extraction out of the sequence data before filtering. Thus, potential loss of relevant information is avoided in the optimization test.

In order to attenuate possible outlying results caused by accidental initial populations, every ES test has been executed 3 times independently. The mean results from these 3 independent tests are shown as bar chart in figure 6.11, in which the error bars indicate the minimal and maximal results. It can be concluded from these results that the min_max and the erf_int target functions yield high precision and good recall rates and that their performance differs only slightly between independent test runs (which is shown by the error bars).

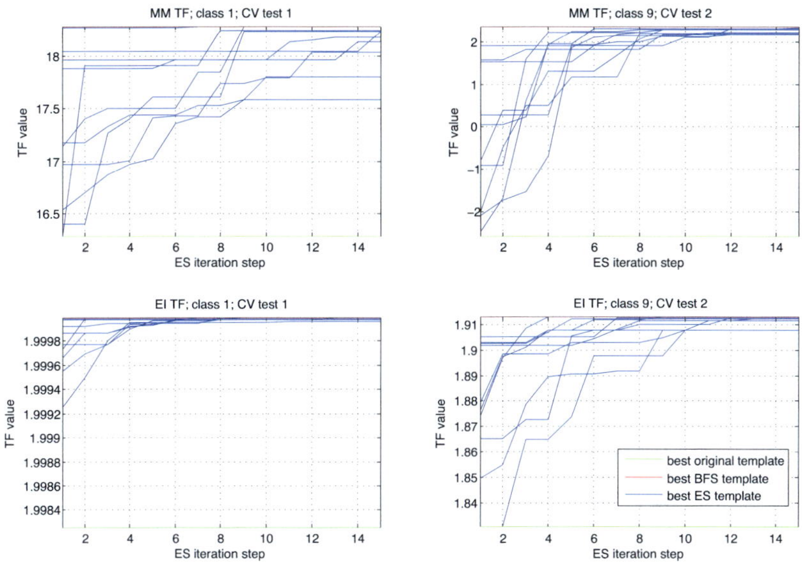

Figure 6.10: Progress of the target function values in several tests with different target functions and classes (blue lines). The results for the best original template are shown by the green lines and the results for the best BFS prototype by the red lines.

Figure 6.12 shows a comparison of ES target function results to a test with prototypes chosen from original templates. The test results are averaged over all classes and participants. Original template prototypes have been chosen by selecting the one with the smallest average distance to all other templates of the same class in the training set. This common method for choosing prototypes is referred to as minimum selection in [54]. Thresholds for minimum selection have been determined by taking the average distance plus two standard deviations. In contrast to ES tests, adding a tolerance to labels for template extraction is not advisable here, because it rather corrupts relevant template information than being of advantage (this has been confirmed in other tests, which are not shown here).

In the figure it can be seen that precision is improved by utilizing templates optimized with the min_max, the KL_div or the erf_int target function. Recall rates have been improved with the min_max and the mean_div target

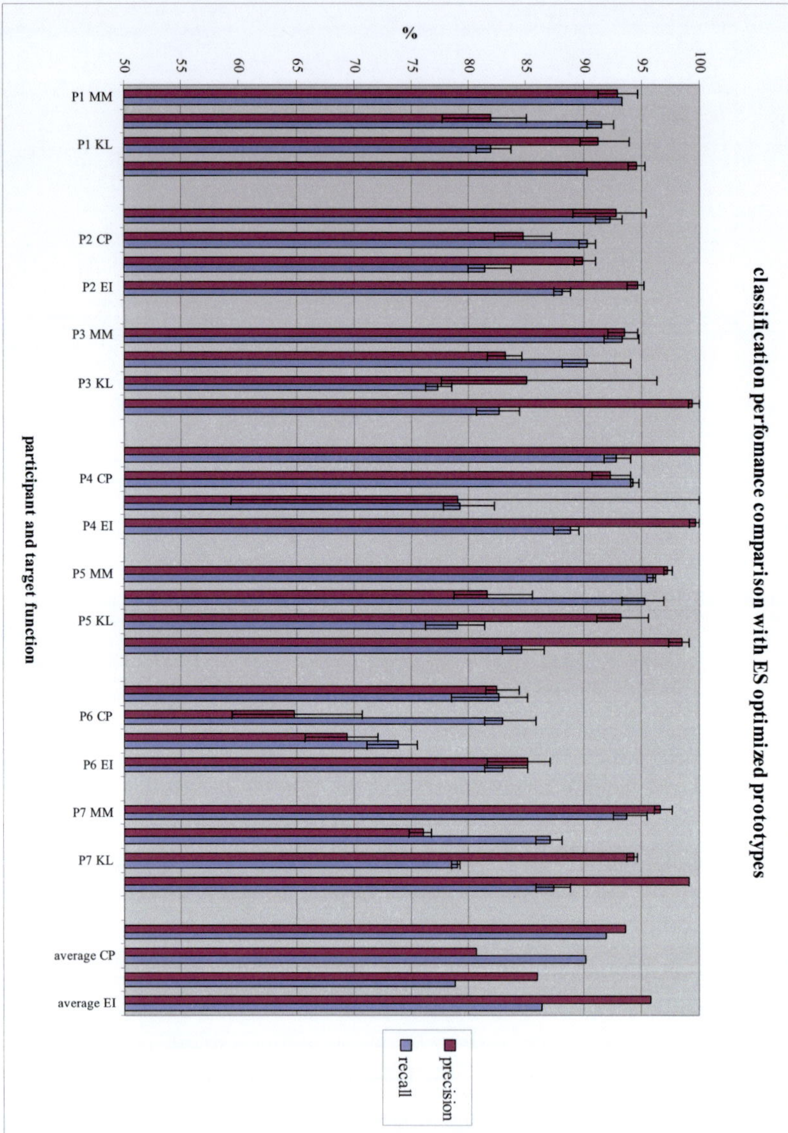

Figure 6.11: Online DTW classification results of a target function performance comparison test with ES optimized prototypes. Results are averaged over all classes and shown for each participant (P1-P7) and target function (MM = min_max, CP = CP_dist, KL = KL_div, EI = erf_int). The error bars indicate maximal and minimal results of 3 independent test runs.

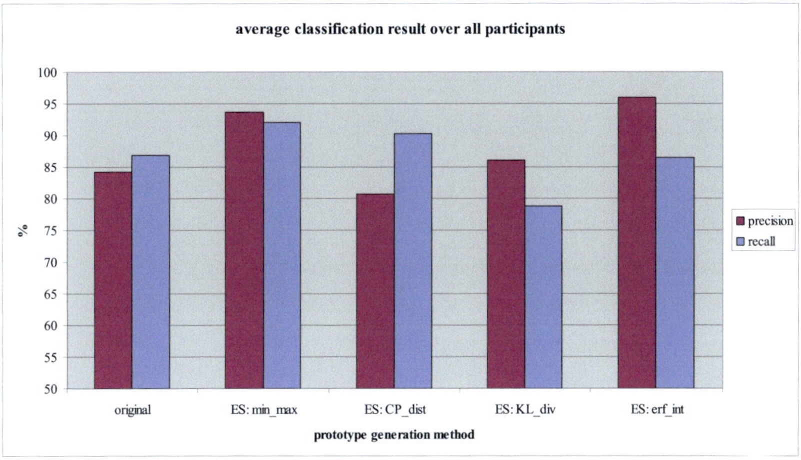

Figure 6.12: Comparison of online DTW results from prototypes optimized with the ES and (original) prototypes chosen by minimum selection. Results are averaged over all participants and classes.

function, while an equal result is achieved with the erf_int target function. The experiments have shown that prototypes optimized with the min_max and the erf_int target functions yield better results than original prototypes (with clear improvements especially for precision rates).

6.5.3. DTW Gesture Recognition Results

In this section, the performance of our online DTW implementation for gesture recognition is evaluated based on tests performed with the test dataset. In the following experiments, the usefulness of range normalization and ambiguity resolution are shown. Furthermore, results from a comparison test of online DTW to gesture recognition with common activity recognition methods are presented.

Effect of range normalization on DTW results

In order to have an account of the effect of the range normalization method, a comparison test has been conducted, in which online DTW with and without range normalization is compared. For this test original template prototypes

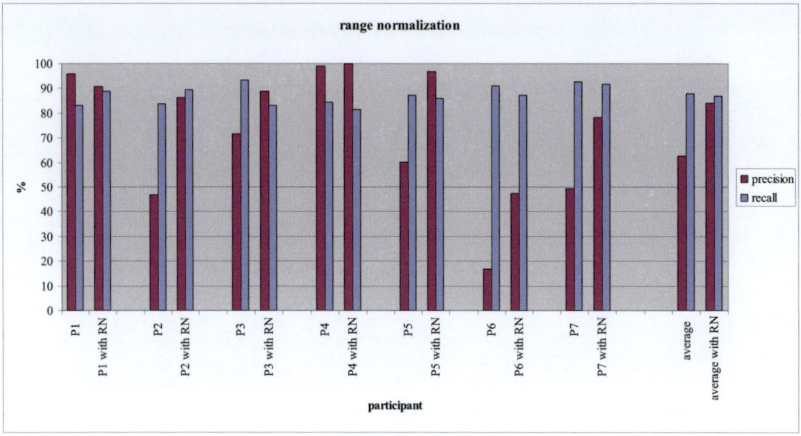

Figure 6.13: Online DTW test results with and without range normalization (RN) utilizing original templates. The test shows the influence of range normalization on the classification result. Results are averaged over all classes.

have been utilized, which have been generated by minimum selection as explained in the previous section.

The test results for each participant are shown in figure 6.13. While range normalization has only a small influence on recall with no clear tendency of improvement, the test has shown that the application of range normalization yields significantly higher precision values for most of the participants. Averaged over all participants the utilization of range normalization has brought an improvement of 20 percent to precision.

Online DTW with ambiguity resolution

Due to the parallel execution of DTW, online DTW results may contain overlapping detections if more than one prototype is used. Therefore, ambiguity resolution (section 6.2.4) has to be applied in order to obtain the final online gesture recognition result in case if more than one prototype is used.

Figure 6.14 shows an online DTW result before and after ambiguity resolution. In this test the online DTW result from the previous section has been used, which has been produced by utilizing prototypes that have been optimized with the ES and the erf_int target function. Due to the fact that few ambiguities existed in the online DTW result because of the good recogni-

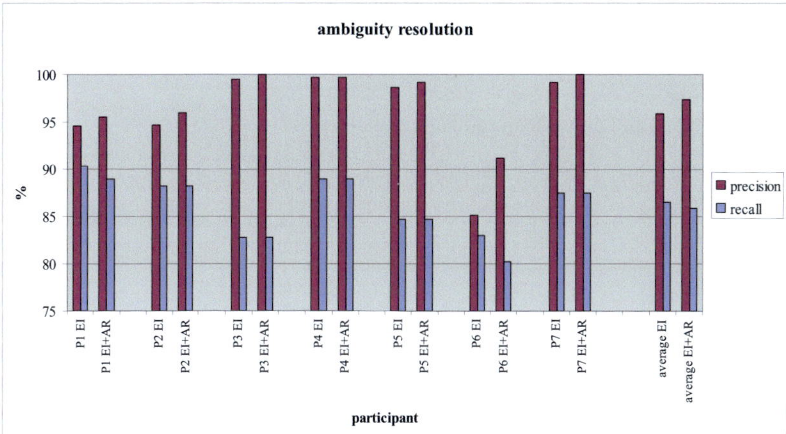

Figure 6.14: Results from online DTW utilizing prototypes optimized with the ES and the erf_int target function before and after the application of ambiguity resolution (AR). Results are averaged over all classes.

tion performance, only slightly different recognition rates resulted in this test after ambiguity resolution. However, in the figure it can be seen that ambiguity resolution leads to an increase of precision and a slight decrease of recall. These effects result from the elimination of false positives (increase of precision) and from a few true positives that have falsely been cleared away (decrease of recall).

Comparison of DTW to standard classifiers

Finally, the results of online DTW have been compared with the sliding window based classification methods that have been utilized for action recognition in section 4.2. This test has been performed in order to prove that the template matching approach with online DTW is more useful for gesture recognition than the generalizing sliding window classification approach.
Figure 6.15 shows the results of a k-nearest neighbor (k-NN) and a Naive Bayes classifier in comparison with the DTW online result. The DTW results in the figure are the same as in the previous test (in which templates have been generated with the ES and the erf_int target function and ambiguity resolution has been utilized). Classifier results have been produced by utilizing PCA feature extraction, a window size of 1024 sampling points (or

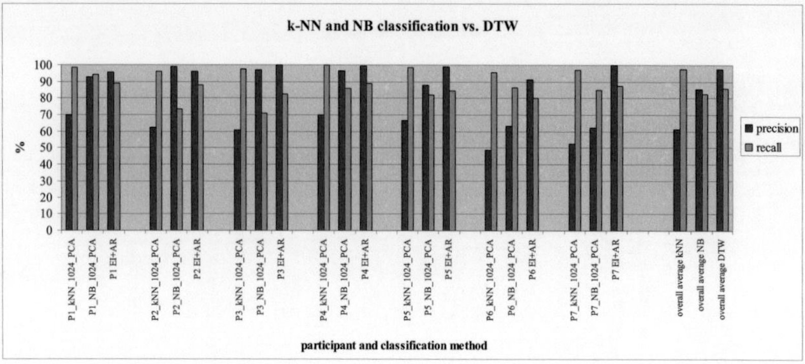

Figure 6.15: Comparison of the classification performance of online DTW with sliding window classification methods. Sliding window classification has been performed with a Naive Bayes (NB) classifier and a k-nearest neighbor classifier (kNN). Results are averaged over all classes.

2.5 s) and a window spacing of 512 sampling points (or 1.25 s). For k-NN classification k has been set to k=3.
Although the k-NN classifier yields a very good recall performance it produces many false positives, which results in poor precision rates. In comparison with the Naive Bayes classifier result, online DTW has better precision rates in most single participant tests and yields a better performance on average (+11.8% precision and +3.2 recall%).

6.6. Summary

The gesture recognition system, which has been presented in this chapter, constitutes a solution that enables the worker to communicate with a superordinate framework at his workplace in a comfortable way via gestures. Because of the fact that the system is based on accelerometers, it is more robust to disturbances than systems based on remote sensing techniques (such as cameras). The small size of accelerometer devices makes them integrable in a working glove or a bracelet.

Gesture recognition and prototype optimization

For the recognition of gestures in streams of sensor data, an online gesture recognition method has been developed that is based on DTW. A problem of the online DTW algorithm is that ambiguities caused by overlapping detections may occur if more than one prototype is used. However, this problem can be treated with the presented ambiguity resolution solution. Since range normalization is applied to templates before distance calculations, spatial variations (which are present in templates that are recorded with accelerometers) are compensated. Thus, the online DTW approach in combination with range normalization is able to deal with gesture templates that vary in length and amplitude.
Time series prototypes representing classes of gestures are generated in our system by means of a novel prototype optimization approach. In this approach, the task of finding prototypes has been formulated as an optimization problem, which aims at the maximization of class separability of time series prototypes. The optimization problem has been described with four different target functions. Optimization techniques for searching the target space have been proposed in form of a brute force search and an evolution strategy. By defining the target space as the set of all subseries of the original time series templates, the problem of inaccurately labeled start and endpoints can be circumvented.

Performance

Although optimization with the BFS guarantees to find the target function maximum, the BFS is rather not applicable to most template datasets, because of the massive consumption of computing time. In performed comparison tests, the ES has proven to be much more efficient for prototype optimization, since it is able to find good solutions in reasonable time.
The result of the comparison test with the different target functions has shown that the erf_int target function yields good classification results and the best precision rates. For our application, high precision rates are even more desirable than high recall rates, because repeating a gesture is not as inconvenient as commands that are issued unwittingly by false or confused detections. The

classification performance of prototypes optimized with the min_max target function is satisfying as well. Prototypes from both target functions yield clearly better results than original templates chosen by minimum selection. Classification with prototypes from the CP_dist and the Kl_div target functions show no clear improvement. The CP_dist target function seems to be a too imprecise measure for class separability and the Kullback-Leibler divergence is basically a good measure for dissimilarity but this must not be the same for class separability.

As expected, the positive effect of range normalization on online DTW has been proven to be useful for gestures recorded with accelerometers and shows a strong improvement of the precision rate.

The application of ambiguity resolution to a multiclass DTW problem has shown that FP's overlapping with TP's are usually cleared away, which has a positive effect on precision. However, occasionally the opposite case occurs and recall rates are slightly reduced.

The comparison test of online DTW with windowed classification methods has shown that DTW yields better results for gesture recognition than windowed classification. This result underlines the assumption that regarding the temporal signal content is of particular importance for gesture recognition.

In summary, the tests have shown that the presented gesture recognition system works very precisely and yields good recognition rates. When utilizing prototypes optimized with the ES and the erf_int target function, the overall (averaged over the 9 classes and the 7 participants) recognition performance for online DTW after ambiguity resolution is 85.86 % recall and 97.35 % precision.

7. Realtime Experiments

In this chapter the functionality of the developed human-machine interface approach is demonstrated as a complete system for online activity recognition in realtime experiments. The experiments have been conducted with two demonstrators, one for the spot welding scenario and another one for the PC assembly scenario and hand gesture recognition. Both demonstrators are based on the human-machine interface solution from section 2.4 and utilize the components for the recognition of human activities, tasks and gestures that have been presented in the previous chapters. The methods of the human-machine interface components that have been used in the experiments have been chosen depending on the respective scenario.

7.1. Spot Welding Experiment

The spot welding experiment has been conducted with the spot welding scenario, which has briefly been introduced in section 2.2. In the following, a detailed description of the scenario and the demonstrator is given and results of the realtime experiment are presented. The presented spot welding experiment can also be found in our publication [38].

7.1.1. Spot Welding Scenario

The spot welding scenario constitutes an environment for demonstrating the recognition of human worker activities of a handling task under realistic industrial conditions and finds application within the scope of XPRESS.

In the spot welding scenario a human worker performs spot welding tasks with the hand welding gun shown at the bottom of figure 7.1. The workpiece in this scenario is represented by the car door, which is depicted at the top of figure 7.1. For the execution of the welding task it is assumed that the car door is clamped in a fixture and that the positions of the welding spots are known.

The chosen demonstration task comprises 4 welds at defined positions (indicated as welding spots in figure 7.1). The welding gun has to be removed from a storage position at the beginning of the task and is returned to it when the spots have been welded. For modelling the worker task with a state-based approach, the following states have been identified:

1. Gun in storage: It is assumed that the welding gun is in a storage position at the beginning of each task. This state is indicated by a storage switch at the welding gun.

2. Tool moving: After picking up the welding gun from the storage position or after the execution of a weld, the worker is moving the welding gun in the work environment.

3. Aligned to spot number n: Before executing a weld, the worker has to move the welding gun tip to a spot position.

4. Ready to weld spot number n: After moving to a spot position, the welding gun has to be kept still in order to ensure that a weld is of an acceptable quality. This means that remaining motion (e.g. shaking) should be minimized.

5. Welding spot number n: A weld is executed by pulling the welding gun trigger.

6. Task completion: The task is considered as completed if all 4 spots have been welded and if the welding gun has been returned to the storage position.

7. Task error: Besides the recognition of the welding task it is also desired to detect and avoid the occurrence of task execution errors. A task execution error occurs when the worker tries to weld at a wrong position or at the position of an already welded spot. Furthermore, it has to be avoided that the welding gun is still in motion when a weld is executed.

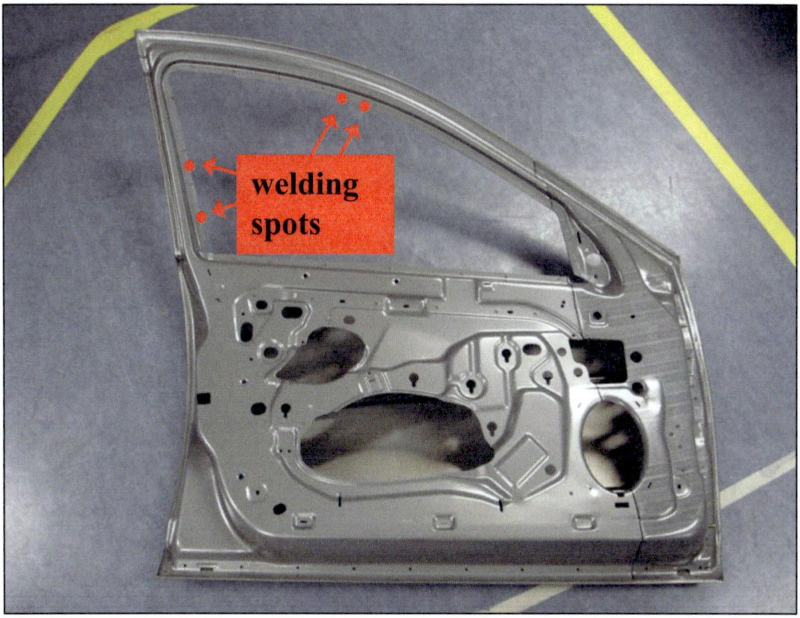

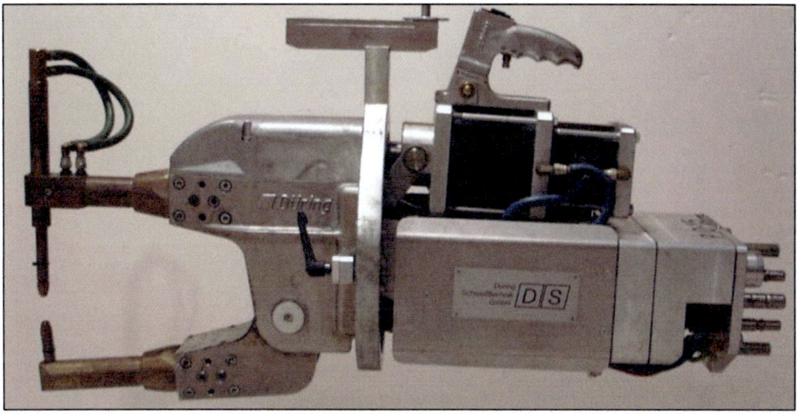

Figure 7.1: Car door (top) and hand welding gun (bottom).

7.1.2. Spot Welding Demonstrator

The experimental setup of the spot welding demonstrator is illustrated in figure 7.2. For robust tracking of the welding gun tip position with the VT

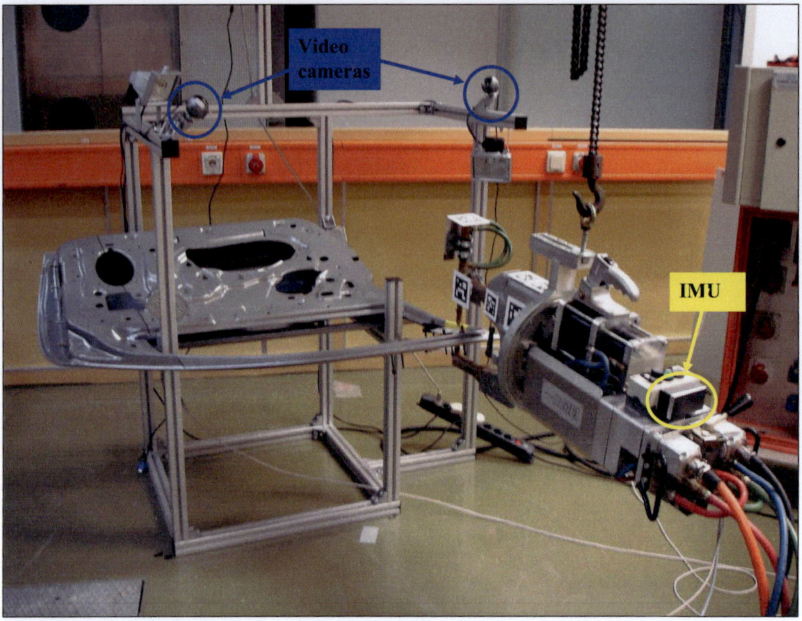

Figure 7.2: Experimental setup of the spot welding demonstrator with welding gun and car door. The locations of the video cameras and the IMU are indicated.

system, several fiducial markers have been attached to the welding gun and two (low cost) video cameras have been installed for observation of the work environment. Additionally, an IMU has been attached to the welding gun. By means of the IMU, welding gun motions can be detected more accurately and faster than with the VT system. The application of a KF is not necessary for this scenario, since occlusions and fast motions do not occur because of the size of the tool. Furthermore, tool state information about the welding gun is available in the form of the welding gun trigger and the storage position switch.

Based on the sensor components and the VT system, locations and actions are recognized. For the classification of spot locations, the deviation of the welding gun tip position $\vec{p}_{VT}$ from a spot reference position $\vec{p}_{ref,n}$ is needed:

$$d_{spot,n} = ||\vec{p}_{VT} - \vec{p}_{ref,n}||. \tag{7.1}$$

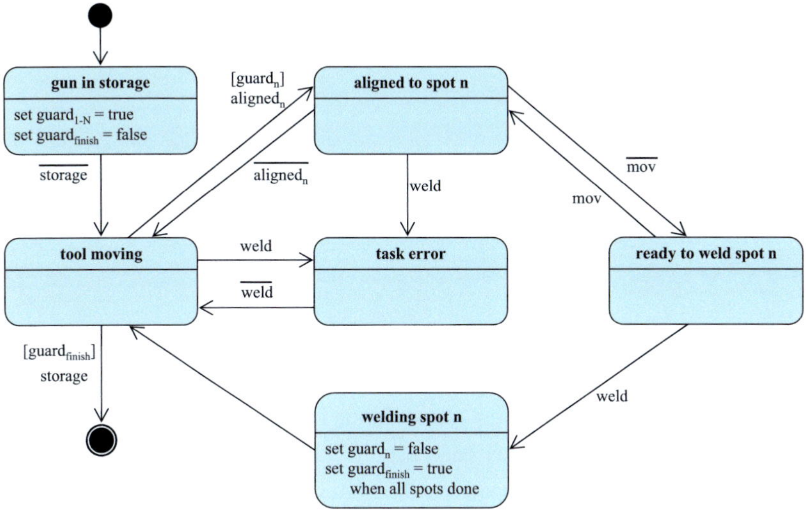

Figure 7.3: Statechart model of the spot welding task. Transitions between task states are modeled as actions and locations (storage = storage switch, weld = welding gun trigger, aligned$_n$ = below position threshold of spot n, mov = above motion threshold).

Because of the fact that reference positions are exactly known in this scenario, thresholding of the spot deviations is sufficient for deriving location information. For the experiment, the threshold has been set to allow a deviation tolerance of 15mm around each welding spot. Information about performed actions is obtained directly from the welding gun trigger and the storage switch. Additionally, motions of the welding gun are recognized by thresholding of the absolute value of measured accelerations $||\vec{a}||$ and measured angular rates $\vec{\omega}$.

Since the spot welding scenario allows to derive location and action information that is not affected with uncertainty, a statechart model has been utilized for modelling the welding task. The statechart model of the spot welding task described in the previous section is given in figure 7.3. As shown in the figure, transitions between task states are represented by action and location changes. Furthermore, guard conditions are used to model the condition that every spot should be welded once (if a guard condition is set to “false”, the respective transition is inhibited).

For the spot welding demonstrator all above mentioned software components have been implemented in C++.

7.1.3. Experiment

The spot welding experiment contains two tests, in which the recognition of the spot welding task has been demonstrated in realtime. For both tests a mockup of a welding gun has been used. The reason for this was to avoid wear of workpiece material and welding equipment. However, it has to be noted that the mockup has similar dimensions as the real welding gun and that the results with the mockup are exactly the same as in official project demonstrations, in which the real welding gun has been used.
In the first test, the worker has welded the 4 welding spots after another in a correct manner. The interface outputs that have been recorded during the execution of the task are shown in figure 7.4. Disregarding a few temporal inconsistencies that result from the manually recorded GT, it can be seen in the figure that the statechart states match with the GT. This shows that the task has been recognized successfully.
The second test exemplifies the detection of task errors. For this test the worker executed the spot welding task in an incorrect manner, as shown in figure 7.5. The incorrect task execution shown in the plot started with a correct weld of spot number 2. After that, the worker pulled the welding gun trigger without having aligned the welding gun tip to a welding spot, which caused the first task error. The second task error occurred when the worker aligned the welding gun to spot number 3 and pulled the welding gun trigger while the welding gun was shaking. Finally, the worker aligned the welding gun to the already welded spot number 2 and pulled the welding gun trigger. In demonstrations with the real welding gun, the execution of a weld was inhibited if a task error had been detected. Thus, the incorrect execution of a task can be avoided when task errors occur.

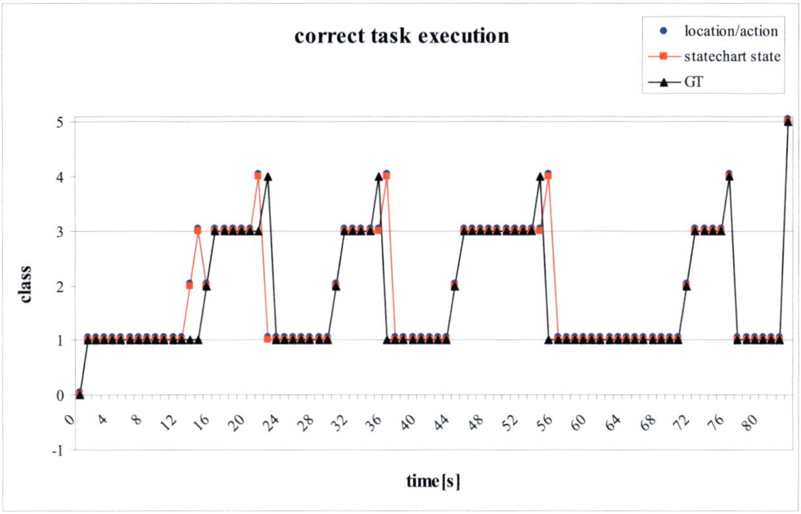

Figure 7.4: Recognition results of a correctly executed spot welding task. The plot shows detected actions and locations such as the resulting statechart states (0 = gun in storage, 1 = gun moving, 2 = gun at spot position, 3 = gun stable at spot position, 4 = welding of spot, 5 = returned to storage, -1 = task error). The GT refers to the statechart state and has been recorded manually.

7.2. PC Assembly and Gesture Recognition Experiment

The second realtime experiment has been conducted with the PC assembly scenario, which has been introduced in section 4.5.1. Additionally, the parallel recognition of hand gestures is demonstrated in this experiment. The PC scenario is of a higher complexity than the spot welding scenario and constitutes a study of the XPRESS project for the recognition of manual assembly tasks in a laboratory test environment that is similar to a human workplace in the industry.

The following sections contain a description of the PC assembly demonstrator and the results of the realtime experiment conducted with the demonstrator. Furthermore, an account of the computation costs of the developed system components is given.

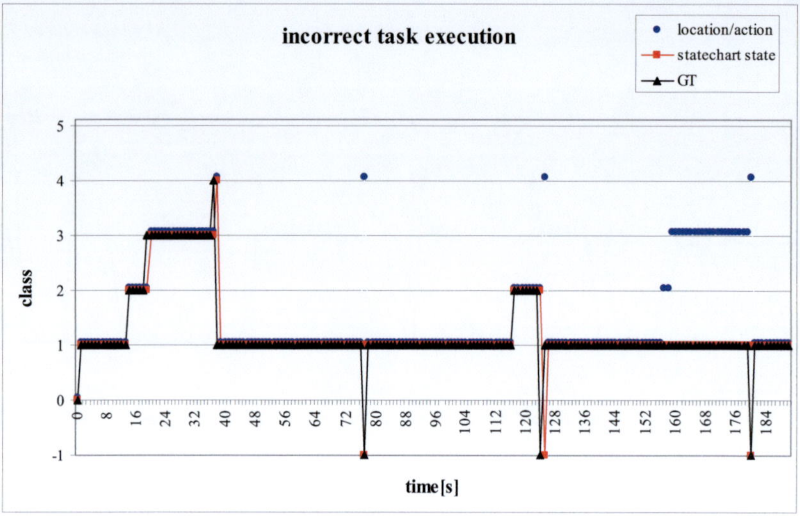

Figure 7.5: Recognition results of an incorrectly executed spot welding task. For the plot the same class indices have been used as in figure 7.4.

7.2.1. PC Assembly Demonstrator

The PC assembly demonstrator utilizes the experimental setup of the scenario description from section 4.5.1. For the demonstrator, the sensor bracelet with the IMU and the industrial cameras are utilized as sensor components. The sensor data is processed with the VT system and the KF in order to provide robust hand position estimations for the location classification component. For action and location classification, the 3-NN classifier and the ML classification methods are used, since this combination proved to be most useful in the PC scenario evaluation. Classified locations and actions are fused by majority voting and are further processed with the (computing time efficient) online HMM in order to recognize activities and tasks.

The recognition of hand gestures is demonstrated with the online DTW approach from chapter 6 and utilizes optimized prototypes of 4 gestures of the gesture recognition dataset from section 6.5.1.

As in the spot welding demonstrator, all software components of the PC assembly demonstrator are implemented in C++.

7.2.2. Experiment

The realtime functionality of the PC assembly demonstrator has been shown in an experiment, in which all 4 assembly tasks have been performed by a test participant while acquired sensor data has been processed with the demonstrator. As a part of the experiment, the 4 different gestures used for the demonstrator have been performed randomly by the participant between and during task execution. Figures 7.6, 7.7, 7.8 and 7.9 show the recognition results for gestures, activities and tasks that have been recorded from the outputs of the human-machine interface while the experiment was running.

From the results of experiment parts 1–4 shown in figures 7.6–7.9 it can be

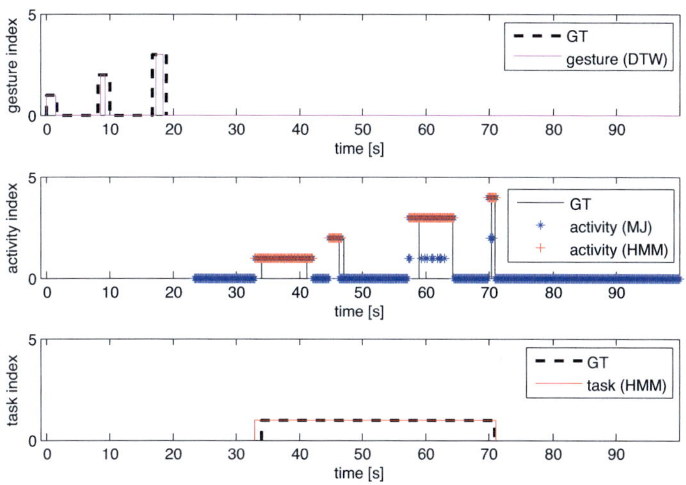

Figure 7.6: Recognition results of the PC assembly and gesture recognition experiment part 1. The results have been recorded during the execution of gestures 1–3 and task number 1.

observed that all detections of the DTW gesture recognition method match with the GT . It can further be observed that the HMM is able to correctly recognize activities and tasks from the uncertain majority voting (MJ) results in figures 7.6–7.8 (occasional temporal inaccuracies result from manually assigned labels and can be neglected). Even in figure 7.9 the performed task could be detected in spite of the poor majority voting result.

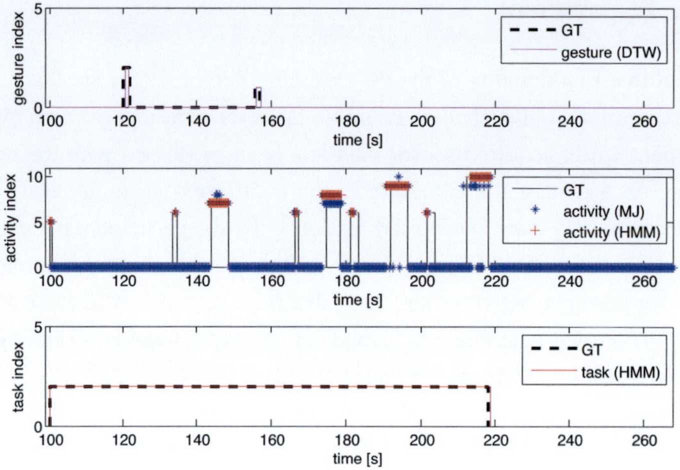

Figure 7.7: Recognition results of the PC assembly and gesture recognition experiment part 2. The results have been recorded during the execution of gestures 2 and 1 and task number 2.

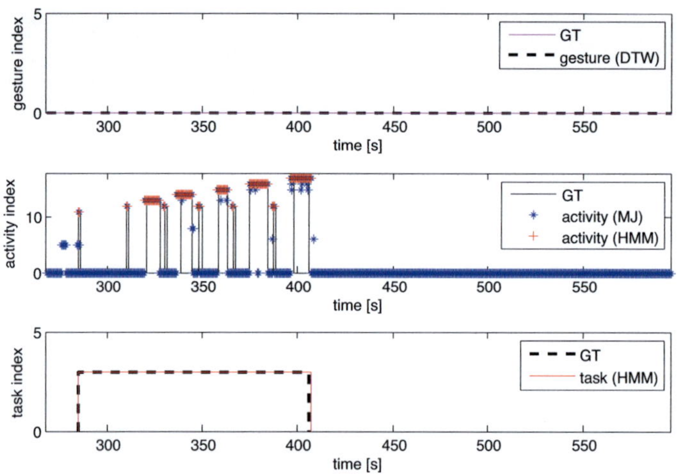

Figure 7.8: Recognition results of the PC assembly and gesture recognition experiment part 3. The results have been recorded during the execution of task number 3.

7.2.3. Computation Costs of Components

In this section an account of the computation costs of the components developed for the PC assembly demonstrator is given. The processing times of

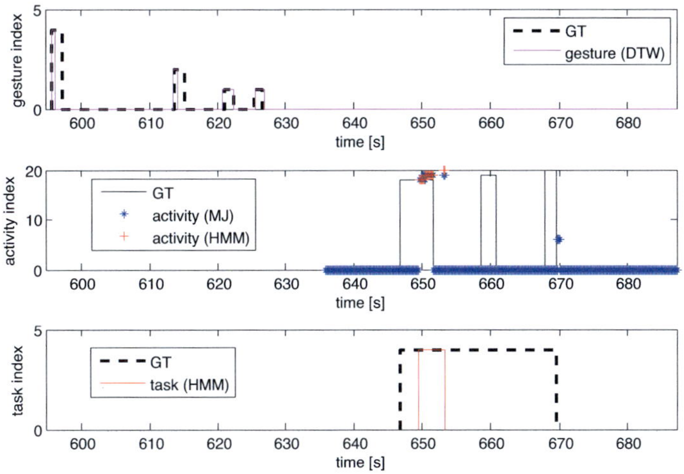

Figure 7.9: Recognition results of the PC assembly and gesture recognition experiment part 4. The results have been recorded during the execution of gestures 4, 2, and 1 (2 times) and task number 4.

the spot welding demonstrator have not been analyzed in particular, since the computation costs of the techniques that differ from the PC assembly demonstrator are comparably low. Figure 7.10 shows a processing time comparison of the software components utilized for the PC assembly demonstrator. The processing times have been measured during the performance of tasks and gestures of the PC assembly scenario. It has to be noted that the VT component is running parallel to the other components of our implementation in order to produce as many position estimations as the available computing resources allow. Therefore, the processing time of the VT component has not been considered in the figure. The processing times have been measured on our test PC (Intel Core 2 Duo E8400 CPU and 2GB DDR2 RAM) with the VT component and the components of the test running on different cores. All processing times listed in the figure have been calculated by taking the average processing time that each component needed to process the input data obtained from the IMU and the VT system during a cycle of one second in reality.

In the figure it can be seen that the whole processing cycle takes less than 50ms for processing all input data measured during 1s.

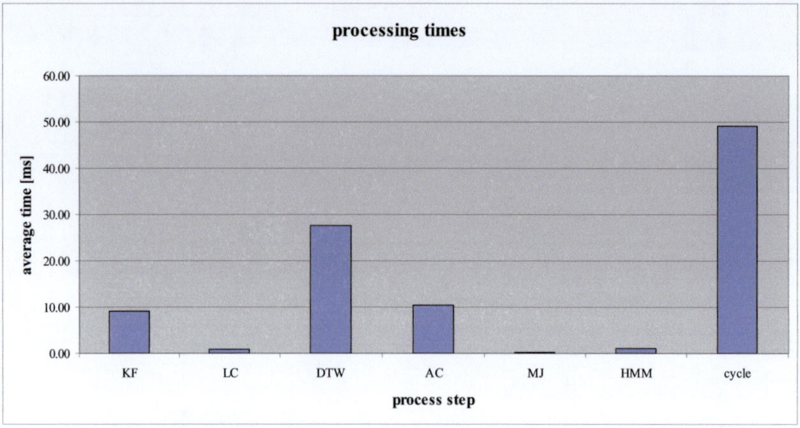

Figure 7.10: Processing time comparison of the components of the PC assembly demonstrator for input data measured during 1s in reality (KF = Kalman filter position classification, LC = location classification, DTW = online DTW gesture recognition, AC = action classification, MJ = activity recognition by majority voting, HMM = online HMM, cycle = complete processing cycle).

7.3. Summary

In this chapter the recognition of activities, tasks and gestures performed by human workers has been demonstrated by means of two demonstrators that are able to process data in realtime.

The first realtime experiment has been conducted with the spot welding demonstrator. By means of the spot welding experiment the recognition of a handling task performed with an industrial tool has been demonstrated. Since the spot welding scenario allows to detect activities that are not affected with uncertainty, spot welding tasks can be modeled in this scenario with the statechart technique. Thus, not only tasks can be recognized but also the detection of task errors is possible.

The second realtime experiment has been conducted with the PC assembly demonstrator, which is able to recognize activities and tasks of a manual assembly scenario, such as gestures for interaction. In the PC assembly and gesture recognition experiment the recognition of tasks and activities from uncertain input information and the recognition of gestures performed paral-

lel to the execution of tasks have been demonstrated in realtime.
Additionally, it has been shown that the human-machine interface is sufficiently fast enough to recognize performed tasks and gestures with a standard PC in realtime.
Since both demonstrators are based on the human-machine interface solution from section 2.4, they have been developed after the same concept. Therefore, it is possible that both demonstrators partially utilize the same components. This demonstrates the flexibility of our worker activity recognition approach with respect to reusability of components.

8. Conclusion

The main objective of this work was to develop a novel human-machine interface for the seamless integration of human workers in software-controlled manufacturing systems. The interface should provide the capability to automatically recognize tasks and activities performed by human workers and should provide a possibility for communication with a superordinate framework via hand gestures. Furthermore, the interface should be flexible with respect to the application by being composed of reusable components, which are able to cover different production scenarios. As a result of this work, an industrial human-machine interface has been developed that provides the above mentioned functionality.

For the development of worker activity recognition systems a conceptual approach has been defined, which describes the worker activity recognition process on 4 levels of abstraction. Based on this concept, a human-machine interface has been developed with components that are not focused on a specific application, but can be utilized in different scenarios. The activity recognition approach of the human-machine interface is based on the general idea to acquire information about what is happening and where something is happening. Thus, both location and action information are incorporated in the activity recognition process. Furthermore, not only low-level activities but also tasks are considered in the worker activity recognition approach, which are modeled as sequences of low-level activities. The concept and the human-machine interface approach provide flexibility with respect to applications, since the defined components are designed to cover different scenarios.

Since the presented activity recognition approach is partially based on location information, a system has been developed that is able to estimate 3D positions of arbitrary objects. The position estimation system utilizes an inertial navigation approach with an extended Kalman filter, which is based on measurements of a small size IMU and marker-based video-tracking. In order to provide measurements with accurate timestamps, the IMU and the utilized industrial cameras have been synchronized in hardware. The navigation approach has been tested and evaluated with real data. Furthermore, the performance of the approach has been analyzed in a simulation, by which it was possible to compare the results with an exactly known reference. In the simulation it has been shown that the accuracy of position estimations is improved by hardware synchronized sensors and the incorporation of bias estimations in the Kalman filter model. In comparison to the VT system alone, the INS solution provides a higher sampling rate and is able to compensate short outages. This allows to capture fast motions and to overcome occlusion problems of short duration. The presented position estimation approach is scalable, since position tracking is possible in dimensions ranging from several meters to a few millimeters.

For the recognition of actions from inertial sensors, a sliding window approach has been utilized. The issue of choosing appropriate features has been regarded by calculation of a variation of features that are commonly used in different activity recognition applications in a first step. In a second step appropriate features are extracted from the features of the first step by application of a principal component analysis. Thus, only relevant features are utilized and the dimensionality of the feature vector is reduced in order to avoid overfitting problems. For the classification of actions from calculated features, a naive Bayes and a k-nearest neighbor classifier have been compared. In tests with real data from a PC assembly scenario it has been shown that a 3-NN yields the best action classification performance. For the classification of locations from position data a density based approach has been proposed. In this approach two different techniques for modelling position densities and classification of locations have been compared, a maximum likelihood method and a nearest neighbor method. An evaluation of the techniques with the PC assembly scenario has shown that the combination of the ML method with Kalman filtered position estimations yields the best recognition results for further processing. Finally, it has been shown that activities can be derived from classified actions and locations by majority voting. This classifier fusion technique has the advantage that it is independent of the clas-

sifier output type and achieves good results in the PC assembly scenario.
In this work not only the recognition of low-level activities, but also the recognition of complex worker behavior in the form of tasks has been considered. The task level activity recognition approach is based on the idea to model tasks as sequences of low-level activities or actions and distinguishes between certain and uncertain input information. In the case of certain input information it is possible to describe worker tasks with a deterministic statechart model. For the recognition of tasks from uncertain input information two probabilistic modelling methods have been compared, a hidden Markov model and a dynamic Bayesian network. Besides the online recognition of tasks, the proposed methods are also able to significantly improve uncertain activity recognition results. The performance of the probabilistic models have been compared in an evaluation with data from the PC scenario, which has been processed with the low-level activity recognition approach. In this experiment it has been shown that both methods are able to recognize tasks from uncertain low-level activity information and to yield high recognition rates that are well above 95% precision and recall. While the DBN yields a higher accuracy for the recognition of the start and the end of a task and better activity recognition results, the HMM is more efficient with respect to the consumption of computing time. Furthermore, it has again been shown that the utilization of Kalman filtered position estimation improves the activity recognition results. By means of the task level activity recognition approach it has been demonstrated that the incorporation of context knowledge is essential for task recognition from uncertain inputs.
The communication capability via hand gestures of the human-machine interface has been realized by utilizing a time series analysis approach that is based on dynamic time warping. For the approach an online dynamic time warping algorithm has been proposed, which is able to recognize multiple gestures in parallel and resolves ambiguous detections. In order to provide gapless gesture recognition results, accelerometer measurements from the IMU are utilized. The important issue of finding representative prototype templates has been tackled with a novel prototype optimization technique, which is based on an evolution strategy. This optimization approach aims at maximizing the class separability of time series templates. In experiments

with a recorded gesture dataset, the performances of different target functions for the mathematical description of the class separability have been compared and an account of the recognition performance of the online gesture recognition system has been given. The proposed approach constitutes a gesture recognition technique that needs comparable few training data, achieves good detection results with an average recall rate of over 85% and works accurately with an average precision rate of more than 97%.

Finally, the developed human-machine interface approach has been demonstrated in realtime experiments with two demonstrators. These demonstrators cover the recognition of a handling task and a manual assembly task in form of a spot welding scenario and the PC assembly scenario. In the spot welding experiment the recognition of the execution of an industrial spot welding task has been demonstrated with the statechart task model and certain input information. Besides that the realtime recognition of tasks and activities from uncertain input information has been demonstrated in an experiment with the PC scenario. Furthermore, the recognition of hand gestures, which are performed parallel to the worker task, have been demonstrated in the PC scenario experiment.

In summary, it has been shown that the developed human-machine interface allows the seamless integration of human workers in an automatically controlled production system. Although the recognition of human behavior constitutes a problem of high complexity, high task and activity recognition rates have been achieved by utilizing task recognition models, which incorporate context knowledge about performed worker activities. Furthermore, not only the recognition of performed tasks but also the possibility for comfortable interactive communication has been provided by means of the gesture recognition system. It has also been demonstrated that the presented approach is able to cover different scenarios and thus constitutes a solution that offers flexibility with respect to the application.

A. Derivation of the Kullback-Leibler Divergence for Normal Distributions

The Kl_div target function that is used in chapter 6 for prototype optimization is based on the symmetric Kullback-Leibler divergence formula assuming underlying normal distributions. In this chapter the symmetric divergence formula for normal distributions is derived.

The symmetric Kullback-Leibler divergence is expressed by the Kullback-Leibler divergence from equation (6.18) according to:

$$d_{1,2} = D_{KL\ 1,2} + D_{KL\ 2,1} = \int_{-\infty}^{\infty} \left(p\left(x|c_1\right) - p\left(x|c_2\right)\right) ln\frac{p\left(x|c_1\right)}{p\left(x|c_2\right)} dx. \tag{A.1}$$

After insertion of the normal distribution formula for $p\left(x|c_1\right)$ and $p\left(x|c_2\right)$ we obtain:

$$\begin{aligned} d_{1,2} &= \int_{-\infty}^{\infty} \left(\frac{1}{\sqrt{2\pi}\sigma_1} e^{-\frac{(x-\mu_1)^2}{2\sigma_1^2}} - \frac{1}{\sqrt{2\pi}\sigma_2} e^{-\frac{(x-\mu_2)^2}{2\sigma_2^2}} \right) \\ &\quad \cdot \left(ln\frac{\sigma_2}{\sigma_1} - \frac{(x-\mu_1)^2}{2\sigma_1^2} + \frac{(x-\mu_2)^2}{2\sigma_2^2} \right) dx. \end{aligned} \tag{A.2}$$

Rearrangement of equation (A.2) leads to:

$$\begin{aligned} d_{1,2} &= ln\frac{\sigma_2}{\sigma_1}\left(\int_{-\infty}^{\infty}\frac{1}{\sqrt{2\pi}\sigma_1}e^{-\frac{(x-\mu_1)^2}{2\sigma_1^2}}dx-\int_{-\infty}^{\infty}\frac{1}{\sqrt{2\pi}\sigma_2}e^{-\frac{(x-\mu_2)^2}{2\sigma_2^2}}dx\right) \\ &+ \int_{-\infty}^{\infty}\left(\frac{1}{\sqrt{2\pi}\sigma_1}e^{-\frac{(x-\mu_1)^2}{2\sigma_1^2}}-\frac{1}{\sqrt{2\pi}\sigma_2}e^{-\frac{(x-\mu_2)^2}{2\sigma_2^2}}\right) \\ &\cdot \left(\frac{(x-\mu_2)^2}{2\sigma_2^2}-\frac{(x-\mu_1)^2}{2\sigma_1^2}\right)dx, \end{aligned} \tag{A.3}$$

in which the term in the first brackets sums up to zero (the integrals in the brackets equal to 1). Then we can write:

$$\begin{aligned} d_{1,2} &= \int_{-\infty}^{\infty}\frac{1}{\sqrt{2\pi}\sigma_1}e^{-\frac{(x-\mu_1)^2}{2\sigma_1^2}}\frac{(x-\mu_2)^2}{2\sigma_2^2}-\int_{-\infty}^{\infty}\frac{1}{\sqrt{2\pi}\sigma_1}e^{-\frac{(x-\mu_1)^2}{2\sigma_1^2}}\frac{(x-\mu_1)^2}{2\sigma_1^2} \\ &- \int_{-\infty}^{\infty}\frac{1}{\sqrt{2\pi}\sigma_2}e^{-\frac{(x-\mu_2)^2}{2\sigma_2^2}}\frac{(x-\mu_2)^2}{2\sigma_2^2}dx+\int_{-\infty}^{\infty}\frac{1}{\sqrt{2\pi}\sigma_2}e^{-\frac{(x-\mu_2)^2}{2\sigma_2^2}}\frac{(x-\mu_1)^2}{2\sigma_1^2}dx. \end{aligned} \tag{A.4}$$

In order to solve the first and the last integral in equation (A.4) we need an auxiliary calculation. At first, we obtain the following solution:

$$\begin{aligned} \frac{(x-\mu_a)^2}{2\sigma_a^2} &= \frac{x^2-2\mu_a x+\mu_a^2}{2\sigma_a^2} \\ &= \frac{x^2-2\mu_a x+\mu_a^2-2\mu_b x+\mu_b^2+2\mu_b x-\mu_b^2}{2\sigma_a^2} \\ &= \frac{(x-\mu_b)^2+2(\mu_b-\mu_a)x+\mu_a^2-\mu_b^2}{2\sigma_a^2}. \end{aligned} \tag{A.5}$$

By insertion of equation (A.5) the first and the last integral expressions of equation (A.4) can be written as:

$$\begin{aligned}
\int_{-\infty}^{\infty} \frac{1}{\sqrt{2\pi}\sigma_a} e^{-\frac{(x-\mu_a)^2}{2\sigma_a^2}} \frac{(x-\mu_b)^2}{2\sigma_b^2} dx &= \frac{1}{2\sigma_b^2} \int_{-\infty}^{\infty} (x-\mu_a)^2 \frac{1}{\sqrt{2\pi}\sigma_a} e^{-\frac{(x-\mu_a)^2}{2\sigma_a^2}} dx \\
&+ \frac{2(\mu_a-\mu_b)}{2\sigma_b^2} \int_{-\infty}^{\infty} x \frac{1}{\sqrt{2\pi}\sigma_a} e^{-\frac{(x-\mu_a)^2}{2\sigma_a^2}} dx \\
&+ \frac{\mu_b^2-\mu_a^2}{2\sigma_b^2} \int_{-\infty}^{\infty} \frac{1}{\sqrt{2\pi}\sigma_a} e^{-\frac{(x-\mu_a)^2}{2\sigma_a^2}} dx \\
&= \frac{\sigma_a^2 + 2(\mu_a-\mu_b)\mu_a + \mu_b^2 - \mu_a^2}{2\sigma_b^2} \\
&= \frac{\sigma_a^2 + (\mu_a-\mu_b)^2}{2\sigma_b^2}. \qquad \text{(A.6)}
\end{aligned}$$

Finally, after insertion of equation (A.6) in equation (A.4) we derive:

$$\begin{aligned}
d_{1,2} &= \frac{\sigma_1^2 + (\mu_1-\mu_2)^2}{2\sigma_2^2} - \frac{1}{2\sigma_1^2}\sigma_1^2 - \frac{1}{2\sigma_2^2}\sigma_2^2 + \frac{\sigma_2^2 + (\mu_2-\mu_1)^2}{2\sigma_1^2} \\
&= \frac{1}{2}\left(\frac{\sigma_2^2}{\sigma_1^2} + \frac{\sigma_1^2}{\sigma_2^2} - 2\right) + \frac{1}{2}(\mu_1-\mu_2)^2 \left(\frac{1}{\sigma_1^2} + \frac{1}{\sigma_2^2}\right) \qquad \text{(A.7)}
\end{aligned}$$

Bibliography

[1] ABDULLA, W.H., D. CHOW and G. SIN: *Cross-words reference template for DTW-based speech recognition systems.* TENCON 2003. Conference on Convergent Technologies for Asia-Pacific Region, 4:1576–1579, Oct. 2003.

[2] ABRAMOWITZ, MILTON and IRENE A. STEGUN: *Handbook of Mathematical Functions With Formulas, Graphs, and Mathematical Tables.* NBS Applied Mathematics Series 55. National Bureau of Standards, Washington, DC, 1964.

[3] AGGARWAL, J. K. and Q. CAI: *Human motion analysis: a review.* Computer Vision and Image Understanding, pages 428–440, 1999.

[4] AHLBERG, J., D. ARSIC, T. GANCHEV, A. LINDERHED, P. MENEZES, S. NTALAMPIRAS, T. OLMA, I. POTAMITIS and J. ROS: *Prometheus: Prediction and interpretation of human behavior based on probabilistic structures and heterogeneous sensors.* European Conference on Artificial Intelligence (ECAI), 2008.

[5] ANTIFAKOS, STAVROS, FLORIAN MICHAHELLES and BERNT SCHIELE: *Proactive Instructions for Furniture Assembly.* In *Proc. Ubicomp 2002, Gothenburg*, pages 351–360. Springer, 2002.

[6] AYERS, DOUGLAS and MUBARAK SHAH: *Monitoring human behavior from video taken in an office environment.* Image and Vision Computing, 19(12):833 – 846, 2001.

[7] BAO, LING and STEPHEN S. INTILLE: *Activity Recognition from User-Annotated Acceleration Data.* Pervasive 2004, pages 1–17, April 2004.

[8] BARDRAM, JAKOB E. and HENRIK B. CHRISTENSEN: *Pervasive Computing Support for Hospitals: An overview of the Activity-Based Computing Project.* Pervasive Computing, IEEE, 6(1):44 –51, 2007.

[9] BARRHO, JÖRG: *Sensor- und bildverarbeitungsgestützte Erkennung von Gefahrensituationen.* PhD Thesis, University of Karlsruhe, Germany, 2007.

[10] BEAUREGARD, S.: *Omnidirectional Pedestrian Navigation for First Responders.* Positioning, Navigation and Communication, 2007. WPNC '07. 4th Workshop on, pages 33 –36, march 2007.

[11] BENBASAT, ARI Y. and JOSEPH A. PARADISO: *An Inertial Measurement Framework for Gesture Recognition and Applications.* In *GW '01: Revised Papers from the International Gesture Workshop on Gesture and Sign Languages in Human-Computer Interaction*, pages 9–20, London, UK, 2002. Springer-Verlag.

[12] BOURKE, A.K., J.V. O'BRIEN and G.M. LYONS: *Evaluation of a thresholdbased tri-axial accelerometer fall detection algorithm.* Gait & Posture, 26(2):194–199, 2007.

[13] BROWN, D.C.: *Close-Range Camera Calibration.* Photogrammetric Engineering, pages 855–866, 1971.

[14] BRUSSEL, HENDRIK VAN: *Holonic Manufacturing Systems, the vision matching the problem.* In *First European Conference on Holonic Manufacturing Systems*, 1994.

[15] BUXTON, HILARY: *Learning and understanding dynamic scene activity: a review.* Image and Vision Computing, 21:125–136, 2003.

[16] CORKE, PETER, JORGE LOBO and JORGE DIAS: *An Introduction to Inertial and Visual Sensing.* International Journal of Robotics Research, 26(6):519–535, 2007.

[17] CORRADINI, ANDREA: *Dynamic Time Warping for Off-Line Recognition of a Small Gesture Vocabulary.* page 82, 2001.

[18] DENG, J.W. and H.T. TSUI: *An HMM-Based Approach for Gesture Segmentation and Recognition.* Pattern Recognition, International Conference on, 3:3683, 2000.

[19] DEY, ANIND K.: *Understanding and Using Context.* Personal and Ubiquitous Computing, 5:4–7, 2001.

[20] DING, HUI, GOCE TRAJCEVSKI, PETER SCHEUERMANN, XIAOYUE WANG and EAMONN KEOGH: *Querying and mining of time series data: experimental comparison of representations and distance measures.* Proc. VLDB Endow., 1(2):1542–1552, 2008.

[21] DUDA, RICHARD O., PETER E. HART and DAVID G. STORK: *Pattern Classification.* Wiley, New York, 2. edition, 2001.

[22] EAGLE, NATHAN and ALEX (SANDY) PENTLAND: *Reality mining: sensing complex social systems.* Personal Ubiquitous Comput., 10(4):255–268, 2006.

[23] EU PROJECT XPRESS: *http://www.xpress-project.eu.*

[24] EULER, STEPHEN: *Grundkurs Spracherkennung.* Friedr. Vieweg und Sohn Verlag, Wiesbaden, 2006.

[25] FERRARIS, FRANCO, GRIMALDI UGO and PARVIS MARCO: *Procedure for Effortless In-Field Calibration of Three-Axis Rate Gyros and Accelerometers.* Sensors and Materials, 7(5):311–330, 1995.

[26] FERREIRA, PEDRO, PAULO AZEVEDO, CÂNDIDA SILVA and RUI BRITO: *Mining Approximate Motifs in Time Series.* Discovery Science, pages 89–101, 2006.

[27] FINK, GERNOT A.: *Mustererkennung mit Markov-Modellen : Theorie, Praxis, Anwendungsgebiete.* Teubner, Stuttgart, 1. edition, 2003.

[28] FOXLIN, E., Y. ALTSHULER, L. NAIMARK and M. HARRINGTON: *FlightTracker: a novel optical/inertial tracker for cockpit enhanced vision.* pages 212 – 221, 2-5 2004.

[29] FOXLIN, E. and L. NAIMARK: *VIS-Tracker: a wearable vision-inertial self-tracker.* Virtual Reality, 2003. Proceedings. IEEE, pages 199 – 206, march 2003.

[30] FRITSCH, J., F. LOMKER, M. WIENECKE and G. SAGERER: *Detecting assembly actions by scene observation.* Image Processing, 2000. Proceedings. 2000 International Conference on, 1:212–215 vol.1, 2000.

[31] FUCHS, ERICH, THIEMO GRUBER, JIRI NITSCHKE and BERNHARD SICK: *On-line motif detection in time series with SwiftMotif.* Pattern Recognition, 42(11):3015 – 3031, 2009.

[32] GAVRILA, D. M.: *The visual analysis of human movement: a survey.* Comput. Vis. Image Underst., 73(1):82–98, 1999.

[33] GHAHRAMANI, ZOUBIN: *Learning Dynamic Bayesian Networks.* Lecture Notes In Computer Science, 1387:168–197, 1997.

[34] GREWAL, MOHINDER S., LAWRENCE R. WEILL and ANGUS P. ANDREWS: *Global Positioning Systems, Inertial Navigation, and Integration.* Wiley-Interscience, 2007.

[35] HAREL, DAVID: *Statecharts: A visual formalism for complex systems.* Sci. Comput. Program., 8(3):231–274, 1987.

[36] HARTMANN, BASTIAN and NORBERT LINK: *Gesture Recognition with Inertial Sensors and Optimized DTW Prototypes.* IEEE International Conference on Systems, Man, and Cybernetics (SMC) 2010, in press, 2010.

[37] HARTMANN, BASTIAN, NORBERT LINK and GERT F. TROMMER. *Indoor 3D position estimation using low-cost inertial sensors and marker-based video-tracking*, pages 319 –326, 4-6 2010.

[38] HARTMANN, BASTIAN, CHRISTOPH SCHAUER and NORBERT LINK: *Worker Behavior Interpretation for Flexible Production.* CESSE 2009: International Conference on Computer, Electrical, and Systems Science, and Engineering, 58:494–502, October 2009.

[39] HARTMANN, BASTIAN, INGO SCHWAB and NORBERT LINK: *Prototype Optimization for Temporarily and Spatially Distorted Time Series.*

AAAI Spring Symposium Series SSS10 "It's All in the Timing: Representing and Reasoning about Time in Interactive Behavior", pages 15–20, 2010.

[40] HOFMANN, FRANK G., PETER HEYER and GÜNTER HOMMEL: *Velocity Profile Based Recognition of Dynamic Gestures with Discrete Hidden Markov Models.* pages 81–95, 1998.

[41] HOL, JEROEN D., THOMAS B. SCHÖN, HENK LUINGE, PER J. SLYCKE and FREDRIK GUSTAFSSON: *Robust real-time tracking by fusing measurements from inertial and vision sensors.* Journal of Real-Time Image Processing, 2(2):149–160, 2007.

[42] HUYNH, DUY TAM GILLES: *Human Activity Recognition with Wearable Sensors.* PhD Thesis, Darmstadt, Germany, 2008.

[43] ITAKURA, F.: *Minimum prediction residual principle applied to speech recognition.* Acoustics, Speech and Signal Processing, IEEE Transactions on, 23(1):67–72, Feb 1975.

[44] JAIN, A.K., R.P.W. DUIN and JIANCHANG MAO: *Statistical pattern recognition: a review.* Pattern Analysis and Machine Intelligence, IEEE Transactions on, 22(1):4–37, Jan 2000.

[45] JAN WENDEL, CHRISTIAN SCHLAILE, GERT F. TROMMER: *Comparison of error state space and total state space Kalman filter for integrated navigation systems.* Proceedings of the 15th annual meeting of the IAR (IAR2000), Nancy, France, 2000.

[46] JENSEN, FINN V. and THOMAS D. NIELSEN: *Bayesian Networks and Decision Graphs.* Springer Publishing Company, Incorporated, 2nd edition, 2007.

[47] JONDRAL, FRIEDRICH and ANNE WIESLER: *Wahrscheinlichkeitsrechnung und stochastische Prozesse: Grundlagen für Ingenieure und Naturwissenschaftler.* 2. Auflage. Teubner B.G. Gmbh, 2002.

[48] JUNKER, HOLGER, OLIVER AMFT, PAUL LUKOWICZ and GERHARD TRÖSTER: *Gesture spotting with body-worn inertial sensors to detect user activities.* Pattern Recognition, 41(6):2010 – 2024, 2008.

[49] KALMAN and RUDOLPH EMIL: *A New Approach to Linear Filtering and Prediction Problems*. Transactions of the ASME–Journal of Basic Engineering, 82(Series D):35–45, 1960.

[50] KAPOOR, A., PICARD R.W.: *Multimodal Affect Recognition in Learning Environments*. ACM MM'05, 2005.

[51] KASTEREN, TIM VAN, ATHANASIOS K. NOULAS, GWENN ENGLEBIENNE and BEN J. A. KRÖSE: *Accurate activity recognition in a home setting*. In *UbiComp*, pages 1–9, 2008.

[52] KATO, HIROKAZU and MARK BILLINGHURST: *Marker Tracking and HMD Calibration for a Video-Based Augmented Reality Conferencing System*. IWAR '99: Proceedings of the 2nd IEEE and ACM International Workshop on Augmented Reality, page 85, 1999.

[53] KELA, JUHA, PANU KORPIPÄÄ, JANI MÄNTYJÄRVI, SANNA KALLIO, GIUSEPPE SAVINO, LUCA JOZZO and SERGIO DI MARCA: *Accelerometer-based gesture control for a design environment*. Personal Ubiquitous Comput., 10(5):285–299, 2006.

[54] KO, MING HSIAO, GEOFF WEST, SVETHA VENKATESH and MOHAN KUMAR: *Using dynamic time warping for online temporal fusion in multisensor systems*. Inf. Fusion, 9(3):370–388, 2008.

[55] KRAUSE, A., D.P. SIEWIOREK, A. SMAILAGIC and J. FARRINGDON: *Unsupervised, dynamic identification of physiological and activity context in wearable computing*. Wearable Computers, 2003. Proceedings. Seventh IEEE International Symposium on, pages 88–97, Oct. 2003.

[56] KULLBACK, S. and R. A. LEIBLER: *On Information and Sufficiency*. The Annals of Mathematical Statistics, 22(1):79–86, 1951.

[57] LAROSE, DANIEL T.: *Data Mining Methods and Models*. John Wiley & Sons, Inc., New York, NY, USA, 2006.

[58] LEE, CHRISTOPHER and YANGSHENG XU: *Online, Interactive Learning of Gestures for Human/Robot Interfaces*. pages 2982–2987, 1996.

[59] LEE, HYEON-KYU and J.H. KIM: *An HMM-based threshold model approach for gesture recognition*. Pattern Analysis and Machine Intelligence, IEEE Transactions on, 21(10):961 –973, oct 1999.

[60] LIANG, RUNG-HUEI and MING OUHYOUNG: *A real-time continuous gesture recognition system for sign language*. In *Automatic Face and Gesture Recognition, 1998. Proceedings. Third IEEE International Conference on*, pages 558 –567, 14-16 1998.

[61] LIU, JIAYANG, ZHEN WANG, LIN ZHONG, JEHAN WICKRAMASURIYA and VENU VASUDEVAN: *uWave: Accelerometer-based Personalized Gesture Recognition and Its Applications*. IEEE PerCom, 2009.

[62] LOMBRISER, CLEMENS, NAGENDRA BHARGAVA BHARATULA, DANIEL ROGGEN and GERHARD TRÖSTER: *On-Body Activity Recognition in a Dynamic Sensor Network*. 2nd International Conference on Body Area Networks, 2007.

[63] LUKOWICZ, P., A. TIMM-GIEL, M. LAWO and O. HERZOG: *WearIT@work: Toward Real-World Industrial Wearable Computing*. Pervasive Computing, IEEE, 6(4):8 –13, 2007.

[64] LUO, R.C., CHIH-CHEN YIH and KUO LAN SU: *Multisensor fusion and integration: approaches, applications, and future research directions*. Sensors Journal, IEEE, 2(2):107–119, Apr 2002.

[65] MINNEN, DAVID, CHARLES L. ISBELL, IRFAN ESSA and THAD STARNER: *Discovering multivariate motifs using subsequence density estimation and greedy mixture learning*. AAAI'07: Proceedings of the 22nd national conference on Artificial intelligence, pages 615–620, 2007.

[66] MOESLUND, THOMAS B., ADRIAN HILTON and VOLKER KRÜGER: *A survey of advances in vision-based human motion capture and analysis*. Comput. Vis. Image Underst., 104(2):90–126, 2006.

[67] MONOSTORI, L., J. VÁNCZA and S.R.T. KUMARA: *Agent-Based Systems for Manufacturing*. CIRP Annals - Manufacturing Technology, 55(2):697 – 720, 2006.

[68] MÖRCHEN, F: *Time Series Knowledge Mining*. PhD Thesis, Marburg, Germany, 2006.

[69] MUEEN, ABDULLAH, EAMONN J. KEOGH, QIANG ZHU, SYDNEY CASH and M. BRANDON WESTOVER: *Exact Discovery of Time Series Motifs*. SDM, pages 473–484, 2009.

[70] NGUYEN, N.T., H.H. BUI, S. VENKATSH and G. WEST: *Recognizing and monitoring high-level behaviors in complex spatial environments*. Computer Vision and Pattern Recognition, 2003. Proceedings. 2003 IEEE Computer Society Conference on, 2:II–620–5, June 2003.

[71] OJEDA, L. and J. BORENSTEIN: *Non-GPS Navigation for Emergency Responders*. 2006 International Joint Topical Meeting: Sharing Solutions for Emergencies and Hazardous Environments, pages 12–15, February 2006.

[72] OLIVER, NURIA, ERIC HORVITZ and ASHUTOSH GARG: *Layered Representations for Human Activity Recognition*. In *ICMI '02: Proceedings of the 4th IEEE International Conference on Multimodal Interfaces*, page 3, Washington, DC, USA, 2002. IEEE Computer Society.

[73] PANTIC, MAJA, ALEX PENTLAND, ANTON NIJHOLT and THOMAS HUANG: *Human Computing and Machine Understanding of Human Behavior: A Survey*. pages 239–248, 2006.

[74] PARNIAN, N. and F. GOLNARAGHI: *A low-cost hybrid SDINS/multi-camera vision system for a hand-held tool positioning*. Position, Location and Navigation Symposium, 2008 IEEE/ION, pages 489 –496, may 2008.

[75] PARNIAN, N. and M.F. GOLNARAGHI: *Integration of vision and inertial sensors for industrial tools tracking*. Sensor Review, 27:132–141(10), 2007.

[76] PENTLAND, ALEX (SANDY): *Automatic mapping and modeling of human networks.* Physica A: Statistical Mechanics and its Applications, 378(1):59–67, May 2007.

[77] POPA, M.C., L.J.M. ROTHKRANTZ, Z. YANG, P. WIGGERS, R. BRASPENNING and C. SHAN: *Analysis of Shopping Behavior based on Surveillance System.* IEEE Int. Conf. on Systems, Man, and Cybernetics (SMC 2010), pages 2512–2519, 2010.

[78] RABINER, LAWRENCE and BIING-HWANG JUANG: *Fundamentals of speech recognition.* PTR Prentice Hall, Englewood Cliffs, NJ, 1993.

[79] RABINER, LAWRENCE R.: *A tutorial on hidden Markov models and selected applications in speech recognition.* pages 267–296, 1990.

[80] RAGHAVENDRA, B.S., C.K. NARAYANAN, G. SITA, A.G. RAMAKRISHNAN and M. SRIGANESH: *Prototype learning methods for online handwriting recognition.* Document Analysis and Recognition, 2005. Proceedings. Eighth International Conference on, 1:287–291, Aug.-1 Sept. 2005.

[81] RAVI, NISHKAM, NIKHIL D, PREETHAM MYSORE and MICHAEL L. LITTMAN: *Activity recognition from accelerometer data.* pages 1541–1546, 2005.

[82] RECHENBERG, INGO: *Evolutionsstrategie. Optimierung technischer Systeme nach Prinzipien der biologischen Evolution.* PhD Thesis, 1973.

[83] REHBINDER, HENRIK and XIAOMING HU: *Drift-free attitude estimation for accelerated rigid bodies.* Automatica, 40(4):653 – 659, 2004.

[84] RETSCHER, G.: *Location Determination in Indoor Environments for Pedestrian Navigation.* pages 547 – 555, april 2006.

[85] ROBERTSON, N. and I. REID: *Behaviour understanding in video: a combined method.* Computer Vision, 2005. ICCV 2005. Tenth IEEE International Conference on, 1:808–815 Vol. 1, Oct. 2005.

[86] RUSSELL, STUART J. and PETER NORVIG: *Artificial Intelligence: a modern approach*. Prentice Hall, 2nd international edition edition, 2003.

[87] RUTA, D. and B. GABRYS: *An Overview of Classifier Fusion Methods*. Computing and Information Systems, 7(1):1–10, 2000.

[88] SAKOE, H. and S. CHIBA: *Dynamic programming algorithm optimization for spoken word recognition*. Acoustics, Speech and Signal Processing, IEEE Transactions on, 26(1):43–49, Feb 1978.

[89] SCHLAILE, CHRISTIAN, OLIVER MEISTER, NATALIE FRIETSCH, CHRISTOPH KESSLER, JAN WENDEL and GERT F. TROMMER: *Using natural features for vision based navigation of an indoor-VTOL MAV*. Aerospace Science and Technology, 13(7):349 – 357, 2009.

[90] SCHLÖMER, THOMAS, BENJAMIN POPPINGA, NIELS HENZE and SUSANNE BOLL: *Gesture recognition with a Wii controller*. TEI '08: Proceedings of the 2nd international conference on Tangible and embedded interaction, pages 11–14, 2008.

[91] SCHÖN, THOMAS and FREDRIK GUSTAFSSON: *Integrated Navigation of Cameras for Augmented Reality*. Proc. 16th IFAC World Congress, July 2005.

[92] SCHWEFEL, HANS-PAUL: *Evolution and Optimum Seeking*. Sixth-Generation Computer Technology. Wiley Interscience, New York, 1995.

[93] SHEN, W., NORRIE D.H.: *Agent-Based Systems for Intelligent Manufacturing: A State-of-the-Art Survey*. Knowledge and Information Systems, an International Journal, pages 129–156, 1999.

[94] SHIN, D., R.A. WYSK and L. ROTHROCK: *An investigation of a human material handler on part flow in automated manufacturing systems*. Systems, Man and Cybernetics, Part A, IEEE Transactions on, 36(1):123–135, Jan. 2006.

[95] SIIRTOLA, P., P. LAURINEN and J. RÖNING: *Mining an Optimal Prototype from a Periodic Time Series: an Evolutionary Computation-based Approach*. Congress on Evolutionary Computation (CEC 2009), pages 2818–2824, May 2009.

[96] STARNER, THAD, ALEX PENTLAND and JOSHUA WEAVER: *Real-Time American Sign Language Recognition Using Desk and Wearable Computer Based Video*. IEEE Trans. Pattern Anal. Mach. Intell., 20(12):1371–1375, 1998.

[97] STIEFMEIER, T., G. OGRIS, H. JUNKER, P. LUKOWICZ and G. TROSTER: *Combining Motion Sensors and Ultrasonic Hands Tracking for Continuous Activity Recognition in a Maintenance Scenario*. Wearable Computers, 2006 10th IEEE International Symposium on, pages 97–104, Oct. 2006.

[98] STIEFMEIER, T., D. ROGGEN and G. TROSTER: *Fusion of String-Matched Templates for Continuous Activity Recognition*. Wearable Computers, 2007 11th IEEE International Symposium on, pages 41–44, Oct. 2007.

[99] STIEFMEIER, T., D. ROGGEN, G. TROSTER, G. OGRIS and P. LUKOWICZ: *Wearable Activity Tracking in Car Manufacturing*. Pervasive Computing, IEEE, 7(2):42–50, April-June 2008.

[100] STIEFMEIER, THOMAS: *Real-Time Spotting of Human Activities in Industrial Environments*. PhD Thesis, ETH Zurich, 2008.

[101] STIEFMEIER, THOMAS, CLEMENS LOMBRISER, DANIEL ROGGEN, HOLGER JUNKER, GEORG OGRIS and GERHARD TRÖSTER: *Event-Based Activity Tracking in Work*. In Proceedings of the 3rd IFAWC, pages 91–100, 2006.

[102] STIKIC, M., T. HUYNH, K. VAN LAERHOVEN and B. SCHIELE: *ADL recognition based on the combination of RFID and accelerometer sensing*. Pervasive Computing Technologies for Healthcare, 2008. PervasiveHealth 2008. Second International Conference on, pages 258 –263, jan. 2008.

[103] SUBRAMANYA, AMARNAG, ALVIN RAJ, JEFF BILMES and DIETER FOX: *Recognizing Activities and Spatial Context Using Wearable Sensors*. Proceedings of the 22nd Annual Conference on Uncertainty in Artificial Intelligence (UAI-06), 2006.

[104] TAO, YAQIN, HUOSHENG HU and HUIYU ZHOU: *Integration of Vision and Inertial Sensors for Home-based Rehabilitation*. Proc. IEEE 2nd Workshop InerVis ICRA, 2005.

[105] TITTERTON, DAVID H and JOHN L WESTON: *Strapdown inertial navigation technology*. Institution of Electrical Engineers, 2nd Edition edition, 2004.

[106] URBAN, M., P. BAJCSY, R. KOOPER and J.-C. LEMENTEC: *Recognition of arm gestures using multiple orientation sensors: repeatability assessment*. Proceedings of the 7th International IEEE Conference on Intelligent Transportation Systems, pages 553–558, Oct. 2004.

[107] URL ANALOG DEVICES : *http://www.analog.com*.

[108] URL CAMERA CALIBRATION TOOLBOX FOR MATLAB : *http://www.vision.caltech.edu/bouguetj/calib_doc/*.

[109] URL THE IMAGING SOURCE : *http://www.theimagingsource.com*.

[110] VAN LAERHOVEN, K. and A.K. ARONSEN: *Memorizing What You Did Last Week: Towards Detailed Actigraphy With A Wearable Sensor*. Distributed Computing Systems Workshops, 2007. ICDCSW '07. 27th International Conference on, pages 47–47, June 2007.

[111] VLASIC, DANIEL, ROLF ADELSBERGER, GIOVANNI VANNUCCI, JOHN BARNWELL, MARKUS GROSS, WOJCIECH MATUSIK and JOVAN POPOVIĆ: *Practical motion capture in everyday surroundings*. ACM Trans. Graph., 26(3):35:1–35:9, 2007.

[112] WAGNER, DANIEL and DIETER SCHMALSTIEG: *ARToolKitPlus for Pose Tracking on Mobile Devices*. Proceedings of 12th Computer Vision Winter Workshop (CVWW'07), 2007.

[113] WANG, LIANG, WEIMING HU and TIENIU TAN: *Recent developments in human motion analysis*. Pattern Recognition, 36(3):585–601, 2003.

[114] WANG, SY BOR, ARIADNA QUATTONI, LOUIS-PHILIPPE MORENCY and DAVID DEMIRDJIAN: *Hidden Conditional Random Fields for Gesture Recognition*. CVPR '06: Proceedings of the 2006 IEEE Computer Society Conference on Computer Vision and Pattern Recognition, pages 1521–1527, 2006.

[115] WARD, JAMIE A., PAUL LUKOWICZ, GERHARD TRÖSTER and THAD STARNER: *Activity recognition of assembly tasks using body-worn microphones and accelerometers*. IEEE Transactions on Pattern Analysis and Machine Intelligence, 28(10):1553–1567, 2006.

[116] WENDEL, JAN: *Integrierte Navigationssysteme: Sensordatenfusion, GPS und Inertiale Navigation*. Oldenbourg, Muenchen, 2007.

[117] WILPON, J. and L. RABINER: *A modified K-means clustering algorithm for use in isolated work recognition*. Acoustics, Speech and Signal Processing, IEEE Transactions on, 33(3):587 – 594, jun 1985.

[118] WILSON, A.D. and A.F. BOBICK: *Realtime online adaptive gesture recognition*. 1:270 –275 vol.1, 2000.

[119] WITTEN, IAN H. and EIBE FRANK: *Data Mining: Practical Machine Learning Tools and Techniques*. Morgan Kaufmann Series in Data Management Sys. Morgan Kaufmann, 2. edition, June 2005.

[120] WOJEK, C., K. NICKEL and R. STIEFELHAGEN: *Activity Recognition and Room-Level Tracking in an Office Environment*. IEEE International Conference on In Multisensor Fusion and Integration for Intelligent Systems, pages 25–30, 2006.

[121] WU, JIANXIN, ADEBOLA OSUNTOGUN, TANZEEM CHOUDHURY, MATTHAI PHILIPOSE and JAMES M. REHG: *A scalable approach to activity recognition based on object use*. In *Proceedings of the International Conference on Computer Vision (ICCV), Rio de*, 2007.

[122] XI, XIAOPENG, EAMONN J. KEOGH, LI WEI and AGENOR MAFRA-NETO: *Finding Motifs in a Database of Shapes*. 2007.

[123] YAMAGUCHI, A., M. OGAWA, T. TAMURA and T. TOGAWA: *Monitoring behavior in the home using positioning sensors*. Volume 4, pages 1977–1979, oct. 1998.

[124] YE, LEXIANG and EAMONN KEOGH: *Time series shapelets: a new primitive for data mining*. KDD '09: Proceedings of the 15th ACM SIGKDD international conference on Knowledge discovery and data mining, pages 947–956, 2009.

[125] YOU, S. and U. NEUMANN: *Fusion of vision and gyro tracking for robust augmented reality registration*. Virtual Reality, 2001. Proceedings. IEEE, pages 71 –78, march 2001.

[126] ZAPPI, P., T. STIEFMEIER, E. FARELLA, D. ROGGEN, L. BENINI and G. TROSTER: *Activity recognition from on-body sensors by classifier fusion: sensor scalability and robustness*. Intelligent Sensors, Sensor Networks and Information, 2007. ISSNIP 2007. 3rd International Conference on, pages 281–286, Dec. 2007.

[127] ZHANG, ZHENGYOU: *Flexible camera calibration by viewing a plane from unknown orientations*. Computer Vision, 1999. The Proceedings of the Seventh IEEE International Conference on, 1:666 –673 vol.1, 1999.

[128] ZHANG, ZHIQIANG, ZHENG WU, JIANG CHEN and JIAN-KANG WU: *Ubiquitous human body motion capture using micro-sensors*. pages 1–5, 2009.